Yamaha XS750 & 850 Triples Owners Workshop Manual

by Mansur Darlington

with additional information on the later 750 and 850 models

by Chris Rogers

Models covered
XS750 2D. 747cc. UK January 1977 to March 1978
XS750 E. 747cc. UK March 1978 to May 1980
XS750 SE. 747cc. UK May 1980 to July 1982
XS750 D and 2D. 747cc. US 1976 to 1977
XS750 E and SE. 747cc. US 1977 to 1978
XS750 F and SF. 747cc. US 1978 to 1979
XS850 G. 826cc. UK May 1980 to April 1985
XS850 G and SG. 826cc. US 1979 to 1980
XS850 H, SH and LH. 826cc. US 1980 to 1981

ISBN 978 0 85696 712 2

© Haynes Group Limited 1994

(340-2T8)

Haynes Group Limited
Haynes North America, Inc

www.haynes.com

Acknowledgements

Our grateful thanks are due to Mitsui Machinery Sales (UK) Limited, who gave permission to use the line drawings used throughout this manual.

Our thanks are also due to Jim Patch of Yeovil Motor Cycle Services, who supplied the Yamaha XS 750 2-D model featured in this manual.

Brian Horsfall gave considerable assistance with the stripdown and rebuilding and devised the ingenious methods for overcoming the lack of service tools. Les Brazier arranged and took the photographs that accompany the text. Jeff Clew edited the text.

Finally, we would also like to thank the Avon Rubber Company, who kindly supplied illustrations and advice about tyre fitting, and NGK Spark plugs (UK) Ltd who furnished advice about spark plug conditions.

About this manual

The author of this manual has the conviction that the only way in which a meaningful and easy to follow text can be written is first to do the work himself, under conditions similar to those found in the average household. As a result, the hands seen in the photographs are those of the author. Unless specially mentioned, and therefore considered essential, Yamaha special service tools have not been used. There is invariably some alternative means of loosening or removing a vital component when service tools are not available, but risk of damage should always be avoided.

Each of the six Chapters is divided into numbered sections. Within these sections are numbered paragraphs. Cross reference throughout the manual is quite straightforward and logical. When reference is made 'See Section 6.10,' it means Section 6, paragraph 10 in the same Chapter. If another Chapter were meant, the reference would read, for example, 'See Chapter 2, Section 6.10'. All the photographs are captioned with a section/paragraph number to which they refer and are relevant to the Chapter text adjacent.

Figures (usually line illustrations) appear in a logical but numerical order, within a given Chapter. Fig. 1.1 therefore refers to the first figure in Chapter 1.

Left-hand and right-hand descriptions of the machines and their components refer to the left and right of a given machine when the rider is seated normally.

Motorcycle manufacturers continually make changes to specifications and recommendations, and these, when notified, are incorporated into our manuals at the earliest opportunity.

We take great pride in the accuracy of information given in this manual, but motorcycle manufacturers make alterations and design changes during the production run of a particular motorcycle of which they do not inform us. No liability can be accepted by the authors or publishers for loss, damage or injury caused by any errors in, or omissions from, the information given.

Introduction to the Yamaha XS 750 Triples

Although the history of Yamaha can be traced back to the year 1887, when a then very small company commenced manufacture of reed organs, it was not until 1954 that the company became interested in motor cycles. As can be imagined, the problems of marketing a motor cycle against a background of musical instruments manufacture were considerable. Some local racing successes and the use of hitherto unknown bright colour schemes helped achieve the desired results and in July 1955 the Yamaha Motor Company was established as a separate entity, employing a work force of less than 100 and turning out, some 300 machines a month.

Competition successes continued and with the advent of tasteful styling that followed Italian trends, Yamaha became established as one of the world's leading motor cycle manufacturers. Part of this success story is the impressive list of Yamaha 'firsts' - a whole string of innovations that include electric starting, pressed steel frame, torque induction and 6 and 8 port engines. There is also the 'Autolube' system of lubrication, in which the engine-driven oil pump is linked to the twist grip throttle, so that lubrication requirements are always in step with engine demands.

Since 1964, Yamaha has gained the World Championship on numerous occasions, in both the 125 cc and 250 cc classes. Indeed, Yamaha has dominated the lightweight classes in international road racing events to such an extent in recent years that several race promoters are now instituting a special type of event in their programme from which Yamaha machines are barred!

In 1970 Yamaha broke into the four-stroke market with the parallel twin 650 cc XS1. This model was soon augmented by a 500 cc model and 750 cc model of similar configuration. The move from two-strokes to four-strokes was a trend followed by the other large Japanese manufacturers, due to the 'energy crises' and world opinion on pollution. The XS 750D was introduced first to America, where it was available in 1976, as the machine to head the Yamaha range, and so compete in the 'Superbike' market position. The unusual parallel three cylinder engine coupled with a still uncommon shaft drive has produced a machine which although later in arrival than comparable models from other manufacturers has immediately enjoyed a good following.

Contents

1977 Yamaha XS 750 2D model

Ordering spare parts

When ordering spare parts for the Yamaha XS 750, it is advisable to deal direct with an official Yamaha agent, who will be able to supply many of the items required ex-stock. Although parts can be ordered from Yamaha direct, it is preferable to route the order via a local agent even if the parts are not available from stock. He is in a better position to specify exactly the parts required and to identify the relevant spare part numbers so that there is less chance of the wrong part being supplied by the manufacturer due to a vague or incomplete description.

When ordering spares, always quote the frame and engine numbers in full, together with any prefixes or suffixes in the form of letters. The frame number is found stamped on the right-hand side of the steering head, in line with the forks. The engine number is stamped on the right-hand side of the upper crank-case, immediately below the right-hand carburettor.

Use only parts of genuine Yamaha manufacture. A few pattern parts are available, sometimes at cheaper prices, but there is no guarantee that they will give such good service as the originals they replace. Retain any worn or broken parts until the replacements have been obtained; they are sometimes needed as a pattern to help identify the correct replacement when design changes have been made during a production run.

Some of the more expendable parts such as spark plugs, bulbs, tyres, oils and greases etc., can be obtained from accessory shops and motor factors, who have convenient opening hours, charge lower prices and can often be found not far from home. It is also possible to obtain parts on a Mail Order basis from a number of specialists who advertise regularly in the motor cycle magazines.

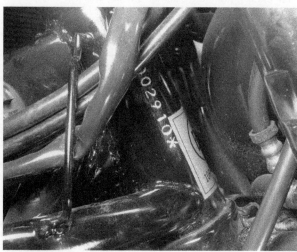

Frame number

Engine number

Safety first!

Professional motor mechanics are trained in safe working procedures. However enthusiastic you may be about getting on with the job in hand, do take the time to ensure that your safety is not put at risk. A moment's lack of attention can result in an accident, as can failure to observe certain elementary precautions.

There will always be new ways of having accidents, and the following points do not pretend to be a comprehensive list of all dangers; they are intended rather to make you aware of the risks and to encourage a safety-conscious approach to all work you carry out on your vehicle.

Essential DOs and DON'Ts

DON'T start the engine without first ascertaining that the transmission is in neutral.

DON'T suddenly remove the filler cap from a hot cooling system – cover it with a cloth and release the pressure gradually first, or you may get scalded by escaping coolant.

DON'T attempt to drain oil until you are sure it has cooled sufficiently to avoid scalding you.

DON'T grasp any part of the engine, exhaust or silencer without first ascertaining that it is sufficiently cool to avoid burning you.

DON'T allow brake fluid or antifreeze to contact the machine's paintwork or plastic components.

DON'T syphon toxic liquids such as fuel, brake fluid or antifreeze by mouth, or allow them to remain on your skin.

DON'T inhale dust – it may be injurious to health (see *Asbestos* heading).

DON'T allow any spilt oil or grease to remain on the floor – wipe it up straight away, before someone slips on it.

DON'T use ill-fitting spanners or other tools which may slip and cause injury.

DON'T attempt to lift a heavy component which may be beyond your capability – get assistance.

DON'T rush to finish a job, or take unverified short cuts.

DON'T allow children or animals in or around an unattended vehicle.

DON'T inflate a tyre to a pressure above the recommended maximum. Apart from overstressing the carcase and wheel rim, in extreme cases the tyre may blow off forcibly.

DO ensure that the machine is supported securely at all times. This is especially important when the machine is blocked up to aid wheel or fork removal.

DO take care when attempting to slacken a stubborn nut or bolt. It is generally better to pull on a spanner, rather than push, so that if slippage occurs you fall away from the machine rather than on to it.

DO wear eye protection when using power tools such as drill, sander, bench grinder etc.

DO use a barrier cream on your hands prior to undertaking dirty jobs – it will protect your skin from infection as well as making the dirt easier to remove afterwards; but make sure your hands aren't left slippery. Note that long-term contact with used engine oil can be a health hazard.

DO keep loose clothing (cuffs, tie etc) and long hair well out of the way of moving mechanical parts.

DO remove rings, wristwatch etc, before working on the vehicle – especially the electrical system.

DO keep your work area tidy – it is only too easy to fall over articles left lying around.

DO exercise caution when compressing springs for removal or installation. Ensure that the tension is applied and released in a controlled manner, using suitable tools which preclude the possibility of the spring escaping violently.

DO ensure that any lifting tackle used has a safe working load rating adequate for the job.

DO get someone to check periodically that all is well, when working alone on the vehicle.

DO carry out work in a logical sequence and check that everything is correctly assembled and tightened afterwards.

DO remember that your vehicle's safety affects that of yourself and others. If in doubt on any point, get specialist advice.

IF, in spite of following these precautions, you are unfortunate enough to injure yourself, seek medical attention as soon as possible.

Asbestos

Certain friction, insulating, sealing, and other products – such as brake linings, clutch linings, gaskets, etc – contain asbestos. *Extreme care must be taken to avoid inhalation of dust from such products since it is hazardous to health.* If in doubt, assume that they *do* contain asbestos.

Fire

Remember at all times that petrol (gasoline) is highly flammable. Never smoke, or have any kind of naked flame around, when working on the vehicle. But the risk does not end there – a spark caused by an electrical short-circuit, by two metal surfaces contacting each other, by careless use of tools, or even by static electricity built up in your body under certain conditions, can ignite petrol vapour, which in a confined space is highly explosive.

Always disconnect the battery earth (ground) terminal before working on any part of the fuel or electrical system, and never risk spilling fuel on to a hot engine or exhaust.

It is recommended that a fire extinguisher of a type suitable for fuel and electrical fires is kept handy in the garage or workplace at all times. Never try to extinguish a fuel or electrical fire with water.

Note: *Any reference to a 'torch' appearing in this manual should always be taken to mean a hand-held battery-operated electric lamp or flashlight. It does **not** mean a welding/gas torch or blowlamp.*

Fumes

Certain fumes are highly toxic and can quickly cause unconsciousness and even death if inhaled to any extent. Petrol (gasoline) vapour comes into this category, as do the vapours from certain solvents such as trichloroethylene. Any draining or pouring of such volatile fluids should be done in a well ventilated area.

When using cleaning fluids and solvents, read the instructions carefully. Never use materials from unmarked containers – they may give off poisonous vapours.

Never run the engine of a motor vehicle in an enclosed space such as a garage. Exhaust fumes contain carbon monoxide which is extremely poisonous; if you need to run the engine, always do so in the open air or at least have the rear of the vehicle outside the workplace.

The battery

Never cause a spark, or allow a naked light, near the vehicle's battery. It will normally be giving off a certain amount of hydrogen gas, which is highly explosive.

Always disconnect the battery earth (ground) terminal before working on the fuel or electrical systems.

If possible, loosen the filler plugs or cover when charging the battery from an external source. Do not charge at an excessive rate or the battery may burst.

Take care when topping up and when carrying the battery. The acid electrolyte, even when diluted, is very corrosive and should not be allowed to contact the eyes or skin.

If you ever need to prepare electrolyte yourself, always add the acid slowly to the water, and never the other way round. Protect against splashes by wearing rubber gloves and goggles.

Mains electricity and electrical equipment

When using an electric power tool, inspection light etc, always ensure that the appliance is correctly connected to its plug and that, where necessary, it is properly earthed (grounded). Do not use such appliances in damp conditions and, again, beware of creating a spark or applying excessive heat in the vicinity of fuel or fuel vapour. Also ensure that the appliances meet the relevant national safety standards.

Ignition HT voltage

A severe electric shock can result from touching certain parts of the ignition system, such as the HT leads, when the engine is running or being cranked, particularly if components are damp or the insulation is defective. Where an electronic ignition system is fitted, the HT voltage is much higher and could prove fatal.

Routine maintenance

Periodic routine maintenance is a continuous process that commences immediately the machine is used and continues until the machine is no longer fit for service. It must be carried out at specified mileage recordings or on a calendar basis if the machine is not used regularly, whichever is the soonest. Maintenance should be regarded as an insurance policy, to help keep the machine in the peak of condition and to ensure long, trouble-free service. It has the additional benefit of giving early warning of any faults that may develop and will act as a safety check, to the obvious advantage of both rider and machine alike.

The various maintenance tasks are described under their respective mileage and calendar headings. Accompanying photos or diagrams are provided, where necessary. It should be remembered that the interval between the various maintenance tasks serves only as a guide. As the machine gets older, is driven hard, or is used under particularly adverse conditions, it is advisable to reduce the period between each check.

For ease of reference each service operation is described in detail under the relevant heading. However, if further general information is required it can be found within the manual in the relevant Chapter.

Although no special tools are required for routine maintenance, a good selection of general workshop tools are essential. Included in the tools must be a range of metric ring or combination spanners, and a set of Allen keys (socket wrenches).

Weekly or every 200 miles

Tyre pressures
1   Check the tyre pressures with a pressure gauge that is known to be accurate. Always check the pressures when the tyres are cold. If the tyres are checked after the machine has travelled a number of miles, the tyres will have become hotter and consequently the pressure will have increased, possibly as much as 8 psi. A false reading will therefore always result.

Tyre pressures:	Solo	
	Front tyre	26 psi (1.8 kg cm^2)
	Rear tyre	28 psi (2.0 kg cm^2)
	Pillion *	
	Front tyre	28 psi (2.0 kg cm^2)
	Rear tyre	33 psi (2.3 kg cm^2)

* *Also for continuous riding at speeds in excess of 60 mph*

Engine/transmission oil level
2   By means of the dipstick fitted to the filler cap in the top of the primary chaincase, check the level of the engine/gearbox oil. If necessary, replenish with SAE 20W/50 engine oil. To check the level accurately, remove and wipe the dipstick. Insert the dipstick so that the filler cap rests on the top of the casing threads. Do not screw the cap inwards. Remove and check the level.

3   *Safety check*
Give the machine a close visual inspection, checking for loose nuts and fittings, frayed control cables etc. Check the tyres for damage, especially splitting on the sidewalls. Remove any stones or other objects caught between the treads. This is particularly important on the front tyre, where rapid deflation due to penetration of the inner tube will almost certainly cause total loss of control.

4   *Legal check*
Ensure that the lights horn and trafficators function correctly, also the speedometer.

Monthly or every 1,000 miles

Complete the tasks listed under the weekly/200 mile heading and then carry out the following checks.

1   *Hydraulic fluid level*
Check the level of the hydraulic fluid in the front brake master cylinder reservoir, on the handlebars, and also the rear brake reservoir, behind the right-hand frame side cover. The level in both reservoirs should lie between the upper and lower level marks. During normal service, it is unlikely that the hydraulic fluid level will fall dramatically, unless a leak has developed in the system. If this occurs, the fault should be remedied AT ONCE. The level will fall slowly as the brake linings wear and the fluid deficiency should be corrected, when required. Always use an hydraulic fluid of DOT 3 or SAE J1703 specification, and if possible do not mix different types of fluid, even if the specifications appear the same. This will preclude the possibility of two incompatible fluids being mixed and the resultant chemical reactions damaging the seals.

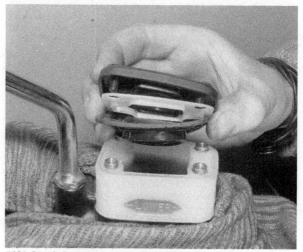

RM1 Check front and rear brake fluid levels

If the level in either reservoir has been allowed to fall below
the specified limit, and air has entered the system, the brake in
question must be bled, as described in Chapter 5, Section 7.

2 Battery electrolyte level

A Yuasa battery is fitted as standard. This battery is a lead-
acid type and has a capacity of 14 amp hours.

The transparent plastic case of the battery permits the upper
and lower levels of the electrolyte to be obersved when the
battery is lifted from its housing below the dualseat. Maintenance is normally limited to keeping the electrolyte level between
the prescribed upper and lower limits and by making sure that
the vent pipe is not blocked. The lead plates and their separators
can be seen through the transparent case, a further guide to the
general condition of the battery.

Unless acid is spilt, as may occur if the machine falls over,
the electrolyte should always be topped up with distilled water,
to restore the correct level. If acid is spilt on any part of the
machine, it should be neutralised with an alkali such as washing
soda and washed away with plenty of water, otherwise serious
corrosion will occur. Top up with sulphuric acid of the correct
specific gravity (1.260 - 1.280) only when spillage has occurred.
Check that the vent pipe is well clear of the frame tubes or any
of the other cycle parts, for obvious reasons.

3 Middle gearcase and final drive box oil levels.

The oil level in both the middle gear case and the final drive
box (at the rear wheel) should be checked at regular intervals.
The filler plug on both units accepts an Allen key, the hexagonal
recess being protected by a removable rubber bung. The oil level
should be checked using the dipstick supplied in the toolkit. Use
the long arm (marked middle) for the middle gear case, and the
short arm (marked rear) for the final drive box. The oil level in
each case should be between the two marked parallel lines. If
required, replenish with Hypoid gear oil as follows.

SAE 90EP for use above $5^{0}C$ ($41^{0}F$)
SAE 80EP for use below $5^{0}C$ ($41^{0}F$)

On US models, Yamaha recommend gear oil conforming to
GL-4, GL-5 or GL-6 specifications. This does not apply to UK
machines, where any good quality EP Hypoid oil will be
satisfactory.

Three monthly or every 3,000 miles

Complete the checks listed under the Weekly/200 mile and
monthly/1,000 mile headings and then carry out the following
tasks:

1 Engine/gearbox oil change

Place a container of more than 3 litre capacity below the
drain plug in the front left-hand wall of the sump. Remove the
filler plug from the primary chaincase and then remove the
drain plug. The oil should be drained when the engine is hot,
preferably after the machine has been on a run, as the lubricant
will be thinner and so drain more rapidly and more completely.

When drainage is complete, refit the drain plug and the
sealing washer and refill the gearbox with approximately 2.8
litres (6/5 US/Imp pints) SAE 20W/50 engine oil. Allow the oil
to settle and then check the level with the filler plug integral
dipstick. When checking the level, do not screw the filler plug
home, but allow it to rest on the edge of the hole.

2 Cleaning and adjusting spark plugs

Remove the three spark plugs and clean them using a wire
brush. Clean the electrode points using emery paper or a fine
file and then reset the gaps. To reset the gap, bend the outer
electrode to bring it closer to or further from the central electrode, until a feeler gauge of the correct size can just be slid
between the gap. Never bend the central electrode or the insulator will crack, causing engine damage if the particles fall in
whilst the engine is running. The correct plug gap is 0.7 - 0.8 mm
(0.028 - 0.032 in). Before replacing the plugs, smear the threads
with a small quantity of graphite grease to aid subsequent
removal.

RM2 Check middle gear oil level, using long dipstick arm

RM3 Check final drive box level using short dipstick arm

RM4 Rotate engine using a large hexagon spanner

RM5 Move fixed point using a screwdriver

RM6 Marks on ATU are for ignition timing

3 *Cleaning and adjusting the contact breaker points*

Remove the contact breaker inspection cover and gasket. The cover is retained by three screws. Inspect the faces of the three sets of contact breaker points. Slight pitting or burning can be removed while the contact breaker unit is in situ on the machine, using a very fine swiss file or emery paper (No. 400) backed by a thin strip of tin. If the pitting or burning is excessive, the contact breaker unit in question should be removed for points dressing or renewal (see Chapter 3, Section 5).

Rotate the engine until one set of points is in the fully open position. The correct gap is within the range 0.3 - 0.4 mm (0.012 - 0.016 in). Adjustment is effected by slackening the screw holding the fixed contact breaker point in position and moving the point either closer or further away with a screwdriver inserted between the small upright post and the slot in the fixed contact plate. Make sure that the points are in the fully open position when this adjustment is made or a false reading will result. When the gap is correct, tighten the screw and recheck.

Repeat the procedure with the other sets of points.

4 *Checking and resetting the ignition timing*

The accuracy of the ignition timing on all three cylinders is vital, if maximum performance and fuel economy is to be maintained. Ignition timing may be checked both manually, with the engine at rest, or by using a stroboscope lamp, with the engine running. The former method, providing care is exercised, is sufficiently accurate, although it has a disadvantage in that the performance of the automatic advance unit (ATU) cannot be checked over the entire rev range. Whichever method is used, the operation should be carried out by referring to the timing marks stamped on the ATU, in conjunction with the index pointer which is secured to the casing behind the contact breaker assemblies base plate. The ATU is stamped with the following marks.

RM7 A = LH cylinder ignition timing adjustment screws

1F   *LH cylinder retarded firing point*
2F   *Centre cylinder retarded firing point*
3F   *RH cylinder retarded firing point*
T   *TDC mark for each cylinder*

In addition, there are three pairs of scribed parallel lines which indicate the full advance position for each cylinder. These lines are unmarked. Commence ignition timing by checking the LH (No. 1) cylinder first. To check the ignition timing manually a small bulb should be connected between the low tension terminal of the contact breaker unit in question and a suitable earth point on the engine. The light will remain on when the points are open and extinguish when the points close. A multimeter connected in a similar manner, with the resistance range

RM8 B = Centre cylinder, C = RH cylinder timing adjustment screws

selected may be substituted in place of the bulb. Switch the
ignition on, (unless using a multi-meter) and using a spanner
applied to the hexagon on the crankshaft end, turn the engine
anti-clockwise. The light should extinguish at the precise point
at which the 1F mark aligns with the index pointer in the casing.
If the timing is incorrect, slacken the three screws which secure
the contact breaker assemblies main baseplate. Rotate the plate
as required until the light flickers and then tighten the screws.
Check that the ignition timing is now correct. Repeat the procedure for the other two cylinders. The top contact breaker
controls the centre cylinder for which the timing mark is 2F.
If the timing is not correct do not move the main baseplate, but
the mounting plate on which the relevant contact breaker is
mounted. Each plate is secured by two screws.

To check the ignition timing, using a stroboscope, interconnect the timing light with the LH cylinder in the manner
described by the manufacturer. Start the engine and allow it to
idle. Point the timing light at the ATU through the aperture in
the contact breaker main baseplate, and check that the 1F mark
aligns with the index pointer. If alignment is not perfect, stop
the engine and adjust the baseplate as described for manual
adjustment. With the retarded timing correct, increase the
engine speed to above 3000 rpm. The index pointer should align
with the small space between the two parallel advance lines for
this cylinder. If the retarded ignition timing is correct but the
advance timing is incorrect, suspect the function of the ATU.
This may be checked as described in Chapter 3, Section 7. The
advance should commence at approximately 1600 rpm. Repeat
the procedure for the centre and RH cylinders.

5   Air filter cleaning

The air filter should be cleaned at the specified routine maintenance interval to ensure an unobstructed flow of air to the
carburettors. Where the conditions are particularly dusty, the
interval should be reduced accordingly.

An air filter cartridge of woven synthetic fibre is fitted in the
air cleaner box to the rear of the carburettors. To gain access to
the element, raise the dualseat and remove the two screws which
secure the air cleaner box cover. Displace the cover and slide the
element up, out of position. The element is supported on a
holder which may be detached by unscrewing the thumb nut.
Tap the element to remove the loose dust and then blow it
through from the **inside**, using an air jet. If the element is damp
or oily, it should be renewed.

6   Carburettor adjustment

The tickover speed of the engine may be altered by turning
the remote adjuster placed between the left-hand and centre
carburettors. The correct idle speed is 1,050 - 1,150 rpm. If
there is any doubt about the carburettor synchronisation, the
carburettors should be checked as described in Chapter 2,
Section 6.14. Refer also to Section 8.

Six monthly or 6,000 miles

Carry out the tasks described in the weekly, monthly and
three monthly sections and then carry out the following:

1   Spark plugs

Remove and renew the spark plugs. Although in general
spark plugs will continue to function after this mileage, their
efficiency will have been reduced. The correct plug type is
NGK BP-7ES or Champion N-7Y. Before fitting, set the gaps to
0.7 - 0.8 mm (0.028 - 0.032 in).

2   Middle gear case and final drive box oil change

Place a container below the rear wall of the gearbox. Remove
the filler plug from the middle gear casing and then unscrew the
drain plug. The drain plug is located in the centre of the middle
gear casing rear wall. Drain the final drive box in a similar manner.
The two cases must be drained when the oil is warm, thereby
ensuring that complete drainage is achieved. The final drive
casing particularly takes an extended length of time to warm up
and therefore should be drained only after a long run. Replenish the two cases with Hypoid gear oil after checking that the
drain plugs have been fitted and tightened. The approximate

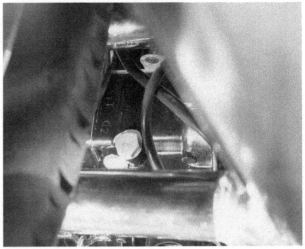

RM9 Middle gear case drain plug

quantities are as follows:

> *Middle gear case 360 cc (12/10 US/Imp fl oz)   EP80/90**
> *Final drive box   300 cc (10/8.5 US/Imp fl oz)   EP80/90**

*see page 15.

Check the levels by means of the dipstick before fitting the
filler plugs, to prevent overfilling.

3   Changing the front fork oil

The damping oil in each fork leg should be drained and
replenished one leg at a time, so that the fork spring remaining
in the undisturbed leg will support the machine.

Place the machine on the centre stand and prise out the
rubber cover from the top of one fork leg. The main spring is
retained by a steel plug secured by a circlip. To remove the
circlip, the plug must be pushed inwards against the pressure of
the fork spring. It is wise to have an assistant available during
this operation as removing the circlip singlehanded whilst
depressing the plug is a difficult process. After removing the
plug, unscrew the drain plug from the lower fork leg and allow
the damping fluid to drain. Pump the forks up and down a
number of times to expel any remaining oil. Replace the drain
plug and replenish the fork leg with 170 cc (5.7/4.8 US/Imp fl oz)
of good quality SAE 20 fork oil or ATF. Refit the plug, circlip
and rubber cover and repeat the procedure for the second fork
leg.

4   Engine oil filter renewal

At this service interval, ie. every second oil change, the engine
oil filter should be renewed. The oil filter, which is of the
corrugated paper type, is contained within a chamber fitted to the
underside of the sump. Access to the filter element is made by
unscrewing the filter chamber centre bolt, which will bring with
it the chamber and element. Place a container below the sump
to catch any oil contained in the chamber.

When renewing the element, it is wise to renew the chamber
'O' ring to prevent possible leaks.

5   Valve clearance adjustment

The clearance between each cam and cam follower is adjusted
by means of removable pads which are seated in the top of each
cam follower. As can be seen by the relevant tables which accompany this section at the beginning of the manual, 25 pads are
available, each of which is of slightly different thickness. Rather
than acquiring the complete set of pads, which in any case may
be duplicated in part by those already in the engine, it is suggested that the required pads be ascertained and purchased, as needed. It is also likely that two or more pads of the same size will be
required.

To gain access to the camshafts, the petrol tank must be
removed and the camshaft cover detached. Drainage of the petrol

tank is not strictly necessary, although doing so at this stage is recommended because the fuel taps require removal for cleaning as described in Section 8.

To check the valve clearance rotate the engine until the cam lobe is pointing away from the cam follower. Insert a feeler gauge and check that the clearance is as follows:

Inlet valves   *0.16 - 0.20 mm (0.006 - 0.008 in)*
Exhaust valves   *0.21 - 0.25 mm (0.008 - 0.010 in)*

If the clearance is incorrect the adjuster pad must be removed to ascertain which pad to install to obtain the correct clearance. A special service tool is available which may be secured to the cylinder head adjacent to the valve in question, and so hold down the cam follower whilst the pad is being removed. The tool can be fitted only when the valve has already been depressed. Rotate the engine until the valve is fully open. Fit the holding tool so that the tongue is contacting the raised edge of the cam follower but not touching the pad. The securing screw must be tightened fully to prevent the tool rotating. Before fitting the tool rotate the cam follower so that the slot in the edge is opposite the tool; this will aid pad removal. The engine must now be turned so that the cam is in the clearance checking position and the pad can be dislodged. It is **absolutely vital** that the cam lobe is **not allowed to touch the holding tool** as the resultant force can break the cylinder head casting. For this reason the exhaust cam must be turned anti-clockwise only, and the inlet cam clockwise only (viewed from the left). Hook out the adjustment pad and note the number stamped on the underside. A magnetic rod or magnetised screwdriver aids this operation considerably.

The correct pad needed to restore the clearance to within the specified range may be found by referring to the relevant table, using the installed pad number and the clearance measured. Select the correct adjuster pad and install it in the cam follower. Rotate the engine until the cam lobe is in contact with the pad and then remove the holding tool. The pad should be positioned with the identification number downwards. Rotate the engine a number of times to ensure that the pad has seated and then recheck the clearance. If the clearance is not satisfactory the adjustment procedure should be repeated using the newly installed pad number as a guide.

If the valve holding tool is not available, an alternative procedure may be adopted, for which an easily constructed cam follower holding tool must be fashioned. The tool may be made from a length of steel plate approximately 6 x ¾ x ¼ in, relieved at one end to fit between the edge of the cam follower and the camshaft. In addition, the tool end must be bent so that it will clear adjacent cylinder head castings. See accompanying photograph.

Rotate the engine until the cam lobe of the valve to be attended to is in the clearance checking position. Using a stout longhandled screwdriver between the cam heel and the adjuster pad, lever the valve into the open position and insert the tool. Ensure that the tool is clear of all castings and securely placed on the edge of the cam follower before releasing the leverage from the screwdriver. Extreme caution should be used when applying this method because if either lever slips, it may result in damage. Use the correct tool if at all possible. After setting the clearances refit the camshaft cover, using a new gasket.

6 *Cylinder head nut and bolt torque check*

Whilst the cylinder head is unobscured by the petrol tank, the torque settings of the various nuts and bolts should be checked. Using the sequence given for tightening in Chapter 1, tighten each nut or bolt to the following settings.

8 mm bolts (Nos. 9, 10)   *2.0 kg m (14 ft lbs)*
10 mm nuts (Nos. 1-8)   *3.5 kg m (25 ft lbs)*

7 *Cam chain tension*

The cam chain is tensioned by a spring-loaded plunger at the rear of the cylinder barrel, on the left-hand side of the machine. Remove the tensioner cap and slacken the plunger detent bolt and locknut. Rotate the engine forwards by applying a spanner to the turning hexagon. As the engine is turned the tensioner plunger will move in and out as the cam chain tension alters. Find the position at which the plunger is at its innermost point. Tighten the detent bolt and locknut and fit the tensioner cap

8 *Fuel filter cleaning*

Before either petrol tap is removed for filter cleaning or for attention to the lever, the petrol tank must be drained. Detach the petrol feed pipes and substitute a longer length of suitable piping through which to drain the fuel. Turn the tap to the 'Prime' position to allow the petrol to flow freely.

To remove either tap, loosen evenly the two screws which pass through the flange into the petrol tank and withdraw the tap, complete with filter tower. The filter is a push fit on the hollow 'T' piece which projects from the top of the tap.

The filter should be cleaned at regular intervals to remove any entrapped matter. Use clean petrol and a soft brush.

9 *Checking the steering head bearings*

Place the machine on the centre stand so that the front wheel is clear of the ground. If necessary, place blocks below the crankcase to prevent the motorcycle from tipping forwards.

Grasp the front fork legs near the wheel spindle and push and pull firmly in a fore and aft direction. If play is evident between the upper and lower steering yokes and the head lug casting, the steering head bearings are in need of adjustment. Imprecise handling or a tendency for the front forks to judder may be caused by this fault.

RM10 Renew filter and check 'O' ring position before replacement

RM11 Check valve/cam clearance

RM12 Yamaha cam follower holding tool

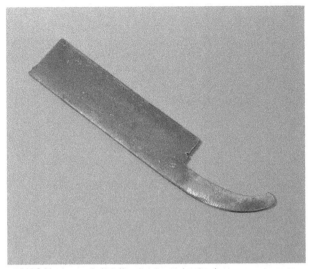

RM13 Home made holding tool may be used to ...

RM14 ... remove valve adjustment pads

RM15 Pad number etched on underside

RM16 Filter is a push fit on 'T' piece

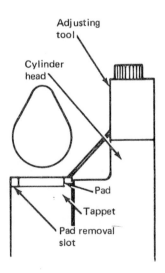

Adjusting valve clearances

Inlet valve adjustment pad selection table

VALVE CLEARANCE (engine cold) 0.16 ~ 0.20mm

Example: Installed is 250
Measured clearance is 0.32mm
Replace 250 pad with 265

*Pad number: (example) Pad No. 250 = 2.50mm
Pad No. 255 = 2.55mm
Always install pad with number down.

MEASURED CLEARANCE (mm)	200	205	210	215	220	225	230	235	240	245	250	255	260	265	270	275	280	285	290	295	300	305	310	315	320
0.00 ~ 0.05				200	205	210	215	220	225	230	235	240	245	250	255	260	265	270	275	280	285	290	295	300	305
0.06 ~ 0.10			200	205	210	215	220	225	230	235	240	245	250	255	260	265	270	275	280	285	290	295	300	305	310
0.11 ~ 0.15		200	205	210	215	220	225	230	235	240	245	250	255	260	265	270	275	280	285	290	295	300	305	310	315
0.16 ~ 0.20																									
0.21 ~ 0.25	205	210	215	220	225	230	235	240	245	250	255	260	265	270	275	280	285	290	295	300	305	310	315	320	
0.26 ~ 0.30	210	215	220	225	230	235	240	245	250	255	260	265	270	275	280	285	290	295	300	305	310	315	320		
0.31 ~ 0.35	215	220	225	230	235	240	245	250	255	260	265	270	275	280	285	290	295	300	305	310	315	320			
0.36 ~ 0.40	220	225	230	235	240	245	250	255	260	265	270	275	280	285	290	295	300	305	310	315	320				
0.41 ~ 0.45	225	230	235	240	245	250	255	260	265	270	275	280	285	290	295	300	305	310	315	320					
0.46 ~ 0.50	230	235	240	245	250	255	260	265	270	275	280	285	290	295	300	305	310	315	320						
0.51 ~ 0.55	235	240	245	250	255	260	265	270	275	280	285	290	295	300	305	310	315	320							
0.56 ~ 0.60	240	245	250	255	260	265	270	275	280	285	290	295	300	305	310	315	320								
0.61 ~ 0.65	245	250	255	260	265	270	275	280	285	290	295	300	305	310	315	320									
0.66 ~ 0.70	250	255	260	265	270	275	280	285	290	295	300	305	310	315	320										
0.71 ~ 0.75	255	260	265	270	275	280	285	290	295	300	305	310	315	320											
0.76 ~ 0.80	260	265	270	275	280	285	290	295	300	305	310	315	320												
0.81 ~ 0.85	265	270	275	280	285	290	295	300	305	310	315	320													
0.86 ~ 0.90	270	275	280	285	290	295	300	305	310	315	320														
0.91 ~ 0.95	275	280	285	290	295	300	305	310	315	320															
0.96 ~ 1.00	280	285	290	295	300	305	310	315	320																
1.01 ~ 1.05	285	290	295	300	305	310	315	320																	
1.06 ~ 1.10	290	295	300	305	310	315	320																		
1.11 ~ 1.15	295	300	305	310	315	320																			
1.16 ~ 1.20	300	305	310	315	320																				
1.21 ~ 1.25	305	310	315	320																					
1.26 ~ 1.30	310	315	320																						
1.31 ~ 1.35	315	320																							
1.36 ~ 1.40	320																								

Exhaust valve adjustment pad selection table

VALVE CLEARANCE (engine cold) 0.21 ~ 0.25mm

Example: Installed is 250
Measured clearance is 0.32mm
Replace 250 pad with 260

*Pad number: (example) Pad No. 250 = 2.50mm
Pad No. 255 = 2.55mm
Always install pad with number down.

MEASURED CLEARANCE (mm)	200	205	210	215	220	225	230	235	240	245	250	255	260	265	270	275	280	285	290	295	300	305	310	315	320
0.00 ~ 0.05					200	205	210	215	220	225	230	235	240	245	250	255	260	265	270	275	280	285	290	295	300
0.06 ~ 0.10				200	205	210	215	220	225	230	235	240	245	250	255	260	265	270	275	280	285	290	295	300	305
0.11 ~ 0.15			200	205	210	215	220	225	230	235	240	245	250	255	260	265	270	275	280	285	290	295	300	305	310
0.16 ~ 0.20		200	205	210	215	220	225	230	235	240	245	250	255	260	265	270	275	280	285	290	295	300	305	310	315
0.21 ~ 0.25																									
0.26 ~ 0.30	205	210	215	220	225	230	235	240	245	250	255	260	265	270	275	280	285	290	295	300	305	310	315	320	
0.31 ~ 0.35	210	215	220	225	230	235	240	245	250	255	260	265	270	275	280	285	290	295	300	305	310	315	320		
0.36 ~ 0.40	215	220	225	230	235	240	245	250	255	260	265	270	275	280	285	290	295	300	305	310	315	320			
0.41 ~ 0.45	220	225	230	235	240	245	250	255	260	265	270	275	280	285	290	295	300	305	310	315	320				
0.46 ~ 0.50	225	230	235	240	245	250	255	260	265	270	275	280	285	290	295	300	305	310	315	320					
0.51 ~ 0.55	230	235	240	245	250	255	260	265	270	275	280	285	290	295	300	305	310	315	320						
0.56 ~ 0.60	235	240	245	250	255	260	265	270	275	280	285	290	295	300	305	310	315	320							
0.61 ~ 0.65	240	245	250	255	260	265	270	275	280	285	290	295	300	305	310	315	320								
0.66 ~ 0.70	245	250	255	260	265	270	275	280	285	290	295	300	305	310	315	320									
0.71 ~ 0.75	250	255	260	265	270	275	280	285	290	295	300	305	310	315	320										
0.76 ~ 0.80	255	260	265	270	275	280	285	290	295	300	305	310	315	320											
0.81 ~ 0.85	260	265	270	275	280	285	290	295	300	305	310	315	320												
0.86 ~ 0.90	265	270	275	280	285	290	295	300	305	310	315	320													
0.91 ~ 0.95	270	275	280	285	290	295	300	305	310	315	320														
0.96 ~ 1.00	275	280	285	290	295	300	305	310	315	320															
1.01 ~ 1.05	280	285	290	295	300	305	310	315	320																
1.06 ~ 1.10	285	290	295	300	305	310	315	320																	
1.11 ~ 1.15	290	295	300	305	310	315	320																		
1.16 ~ 1.20	295	300	305	310	315	320																			
1.21 ~ 1.25	300	305	310	315	320																				
1.26 ~ 1.30	305	310	315	320																					
1.31 ~ 1.35	310	315	320																						
1.36 ~ 1.40	315	320																							
1.41 ~ 1.45	320																								

To adjust the bearings, loosen the pinch bolt which passes through the rear of the upper yoke. Immediately below the upper yoke, on the steering stem, are two peg nuts, the upper being a locknut and the lower the adjuster nut. Using a C spanner loosen the upper nut. Tighten the lower peg nut a little at a time until all play is taken up. Do not overtighten the nut. It is possible to place a pressure of several tons on the head bearings by overtightening even though the handlebars may seem to turn quite freely. Overtight bearings will cause the machine to roll at low speeds and give imprecise steering. Adjustment is correct if there is no play in the bearings and the handlebars swing to full lock either side when the machine is on the centre stand with the front wheel clear of the ground. Only a light tap on each end should cause the handlebars to swing.

Nine monthly or every 9,000 miles

Complete the checks listed under all the previous routine maintenance headings. The following additional tasks are now necessary:

1   *Steering head bearings*
Clean and repack the steering head bearings after dismantling the front forks. Refer to Chapter 4, Sections 2 and 5.

Check the condition of the swinging arm bearings then repack them with grease. See Chapter 4, Section 9 and 10.

Remove, clean and grease the wheel bearings as described in Chapter 5, Sections 8 and 10.

General adjustments and examination

---

1   *Clutch adjustment*
The intervals at which the clutch should be adjusted will depend on the style of riding and the conditions under which the machine is used.

Adjust the clutch in two stages to ensure smooth operation as follows:

Remove the clutch adjustment cover which is a push fit in the left-hand engine cover. Slacken both cable adjuster locknuts and screw fully in both adjusters to gain maximum cable free play. Slacken its locknut and screw in the adjuster screw until

slight resistance is felt; do not overtighten the screw. Slacken the screw by ¼ turn to set the correct amount of free play in the mechanism, then hold it steady while the locknut is tightened. Refit the cover. The cable is adjusted correctly when there is 2 - 3 mm (0.08 - 0.12 in) free play measured between the handlebar lever butt end and the lever clamp (ie 13 - 26 mm/ 0.5 - 1.0 in of free movement at the lever ball end) and the clutch operates correctly with no sign of slip or drag. Use first the lower adjuster to set the free play, reserving the handlebar adjuster for quick roadside adjustments. Tighten the adjuster locknuts and refit the rubber sleeves.

2   *Checking front brake pad wear*
Brake pad wear depends largely on the conditions in which the machine is ridden and at what speed. It is difficult therefore to give precise inspection intervals, but it follows that pad wear should be checked more frequently on a hard ridden machine.

The condition of each pad can be checked easily whilst still in situ on the machine. The pads have a red groove around their outer periphery which can be seen if the small inspection cover in each caliper is lifted. If wear has reduced either or both pads in one caliper down to the red line the pads should be renewed as a pair. In practice, if one set of front pads requires renewal, it will be necessary to renew the other pair. The rear disc brake pads should be checked for wear in the same way as that adopted for the front brakes. The renewal procedure is similar also.

Each set of pads may be removed with the wheel in place by detaching the caliper. There is no need to disconnect the brake hose. Unscrew the two bolts which pass through the fork leg into the caliper support bracket. Lift the assembly off the brake disc and remove the single bolt which holds the caliper unit to the bracket. Unscrew the single crosshead screw from the inner face of the caliper, noting that this screw acts as a locator for the pads. Pull the support bracket from the main unit and lift out the pads. Note the various shims and their positions. The anti-chatter spring fitted between the pad and piston is fitted with the arrow pointing in the direction of wheel rotation.

Fit new pads by reversing the dismantling procedure. If difficulty is encountered when fitting the caliper over the disc, due to the reduced distance between the pads, use a wooden lever to push the piston side pad inwards.

RM17 Check pad wear through inspection window

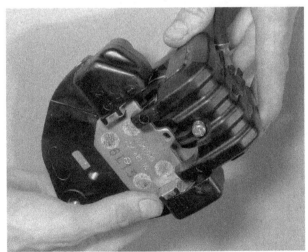
RM18 Detach the caliper from support bracket

Quick glance:
Maintenance adjustments and capacities

Engine oil capacity
Dry ... ... 3.5 litre (7.4/6.1 US/Imp pints)

Oil change ... ... 2.8 litre (6/5 US/Imp pints)

Oil and filter change ... ... 3.2 litre (6.8/5.6 US/Imp pints)

Middle gear case ... ... 360 cc (12/10 US/Imp fl oz)

Final drive box ... ... 300 cc (10/8.5 US/Imp fl oz)

Front forks ... ... 170 cc (5.7/4.8 US/Imp fl oz)

Contact breaker gap ... ... 0.3 - 0.4 mm (0.012 - 0.016 in)

Spark plug gap ... ... 0.7 - 0.8 mm (0.028 - 0.032 in)

Valve clearances (cold)
Inlet ... ... 0.16 - 0.20 mm (0.006 - 0.008 in)
Exhaust ... ... 0.21 - 0.25 mm (0.008 - 0.010 in)

Tyre pressures

	Solo	Pillion
Front	26 psi (1.8 kg cm^2)	28 psi (2.0 kg cm^2)
Rear	28 psi (2.0 kg cm^2)	33 psi (2.3 kg cm^2)

Model dimensions

	XS750D and 2D
Height	1,150 mm (45.3 in)
Width	895 mm (35.3 in)
Length	2,160 mm (85.0 in)
Wheel base	1,470 mm (57.9 in)
Seat height	810 mm (31.9 in)
Weight Dry	232 kg (511 lbs)

Recommended lubricants

Component	Type and specification
Engine and gearbox	SAE 20W/50 in temperatures above 5°C (41°F)
	SAE 10W/30 if temperature does not rise above 15°C (60°F)
Middle gear case	SAE 90 Hypoid gear oil for use above 5°C (41°F)
Final drive box	SAE 80 Hypoid gear oil for use above 5°C (41°F)
Front forks	SAE 20 fork oil or ATF
Wheel bearings	Lithium base high-melting point grease
Hydraulic brakes	DOT 3 (USA) or SAE J1703 brake fluid

Working conditions and tools

When a major overhaul is contemplated, it is important that a clean, well-lit working space is available, equipped with a workbench and vice, and with space for laying out or storing the dismantled assemblies in an orderly manner where they are unlikely to be disturbed. The use of a good workshop will give the satisfaction of work done in comfort and without haste, where there is little chance of the machine being dismantled and reassembled in anything other than clean surroundings. Unfortunately, these ideal working conditions are not always practicable and under these latter circumstances when improvisation is called for, extra care and time will be needed.

The other essential requirement is a comprehensive set of good quality tools. Quality is of prime importance since cheap tools will prove expensive in the long run if they slip or break when in use, causing personal injury or expensive damage to the component being worked on. A good quality tool will last a long time, and more than justify the cost.

For practically all tools, a tool factor is the best source since he will have a very comprehensive range compared with the average garage or accessory shop. Having said that, accessory shops often offer excellent quality tools at discount prices, so it pays to shop around. There are plenty of tools around at reasonable prices, but always aim to purchase items which meet the relevant national safety standards. If in doubt, seek the advice of the shop proprietor or manager before making a purchase.

The basis of any tool kit is a set of open-ended spanners, which can be used on almost any part of the machine to which there is reasonable access. A set of ring spanners makes a useful addition, since they can be used on nuts that are very tight or where access is restricted. Where the cost has to be kept within reasonable bounds, a compromise can be effected with a set of combination spanners – open-ended at one end and having a ring of the same size on the other end. Socket spanners may also be considered a good investment, a basic $3/8$ in or $1/2$ in drive kit comprising a ratchet handle and a small number of socket heads, if money is limited. Additional sockets can be purchased, as and when they are required. Provided they are slim in profile, sockets will reach nuts or bolts that are deeply recessed. When purchasing spanners of any kind, make sure the correct size standard is purchased. Almost all machines manufactured outside the UK and the USA have metric nuts and bolts, whilst those produced in Britain have BSF or BSW sizes. The standard used in USA is AF, which is also found on some of the later British machines. Others tools that should be included in the kit are a range of crosshead screwdrivers, a pair of pliers and a hammer.

When considering the purchase of tools, it should be remembered that by carrying out the work oneself, a large proportion of the normal repair cost, made up by labour charges, will be saved. The economy made on even a minor overhaul will go a long way towards the improvement of a toolkit.

In addition to the basic tool kit, certain additional tools can prove invaluable when they are close to hand, to help speed up a multitude of repetitive jobs. For example, an impact screwdriver will ease the removal of screws that have been tightened by a similar tool, during assembly, without a risk of damaging the screw heads. And, of course, it can be used again to retighten the screws, to ensure an oil or airtight seal results. Circlip pliers have their uses too, since gear pinions, shafts and similar components are frequently retained by circlips that are not too easily displaced by a screwdriver. There are two types of circlip pliers, one for internal and one for external circlips. They may also have straight or right-angled jaws.

One of the most useful of all tools is the torque wrench, a form of spanner that can be adjusted to slip when a measured amount of force is applied to any bolt or nut. Torque wrench settings are given in almost every modern workshop or service manual, where the extent to which a complex component, such as a cylinder head, can be tightened without fear of distortion or leakage. The tightening of bearing caps is yet another example. Overtightening will stretch or even break bolts, necessitating extra work to extract the broken portions.

As may be expected, the more sophisticated the machine, the greater is the number of tools likely to be required if it is to be kept in first class condition by the home mechanic. Unfortunately there are certain jobs which cannot be accomplished successfully without the correct equipment and although there is invariably a specialist who will undertake the work for a fee, the home mechanic will have to dig more deeply in his pocket for the purchase of similar equipment if he does not wish to employ the services of others. Here a word of caution is necessary, since some of these jobs are best left to the expert. Although an electrical multimeter of the AVO type will prove helpful in tracing electrical faults, in inexperienced hands it may irrevocably damage some of the electrical components if a test current is passed through them in the wrong direction. This can apply to the synchronisation of twin or multiple carburettors too, where a certain amount of expertise is needed when setting them up with vacuum gauges. These are, however, exceptions. Some instruments, such as a strobe lamp, are virtually essential when checking the timing of a machine powered by CDI ignition system. In short, do not purchase any of these special items unless you have the experience to use them correctly.

Although this manual shows how components can be removed and replaced without the use of special service tools (unless absolutely essential), it is worthwhile giving consideration to the purchase of the more commonly used tools if the machine is regarded as a long term purchase Whilst the alternative methods suggested will remove and replace parts without risk of damage, the use of the special tools recommended and sold by the manufacturer will invariably save time.

Chapter 1 Engine, clutch and gearbox

Contents

Specifications

Engine

Type	Three cylinder, dohc, air-cooled four stroke
Bore	68 mm
Stroke	68.6 mm
cc	747
Compression ratio	8.5 : 1
bhp	64 @ 7,500 rpm
Maximum torque	6.4 kg m @ 6,000 rpm

Cylinder barrel

Type	Aluminium alloy, cast iron liners
Standard bore	68.00 - 68.02 mm (2.677 - 2.678 in)
Wear limit	68.10 mm (2.681 in)
Taper limit	0.05 mm (0.002 in)
Ovality limit	0.01 mm (0.0004 in)
Cylinder bore/piston clearance	0.050 - 0.055 mm (0.002 - 0.0022 in)
Wear limit	0.1 mm (0.004 in)

Piston and rings

Piston diameter:	
Standard	67.96 mm or 67.97 mm
1st oversize	68.25 mm
2nd oversize	68.50 mm
3rd oversize	68.75 mm
4th oversize	69.00 mm
Ring end gap:	
Top and 2nd ring	0.2 - 0.4 mm (0.008 - 0.016 in)
Oil control ring	0.2 - 0.9 mm (0.008 to 0.035 in)
Side clearance:	
Top ring	0.04 - 0.08 mm (0.0016 - 0.003 in)
2nd ring	0.03 - 0.07 mm (0.0012 - 0.0028 in)

Valves and valve springs

Valve seat angle	45°
Valve stem diameter:	
Inlet	6.975 - 7.010 mm (0.2746 - 0.2759 in)
Exhaust	6.960 - 7.025 mm (0.2740 - 0.2765 in)
Valve guide/stem clearance:	
Inlet:	
Standard	0.020 - 0.041 mm (0.0008 - 0.0016 in
Maximum	0.10 mm (0.004 in)
Exhaust:	
Standard	0.035 - 0.059 mm (0.0014 - 0.0023 in)
Maximum	0.12 mm (0.005 in)
Valve spring free length	
Inner spring	35.6 mm (1.402 in)
Outer spring	39.9 mm (1.571 in)
Valve clearance: engine cold	
Inlet	0.16 - 0.20 mm (0.006 - 0.008 in)
Exhaust	0.21 - 0.25 mm (0.008 - 0.010 in)
Adjuster pad sizes available:	
Pads available in 25 size increments of	0.05 mm (0.002 in) from 2.00 mm (0.080 in)

Camshaft

Inlet:	
Standard lift	8.518 mm (0.3353 in)
Overall lobe height	36.803 $\pm$ 0.05 mm (1.448 $\pm$ 0.002 in)
Wear limit	36.750 mm (1.446 in)
Journal diameter	28.285 $\pm$ 0.05 mm (1.1135 $\pm$ 0.002 in)
Wear limit	28.13 mm (1.107 in)
Exhaust:	
Standard lift	8.018 mm (0.3156 in)
Overall lobe height	36.303 $\pm$ 0.05 mm (1.4292 $\pm$ 0.002 in)
Wear limit	36.15 mm (1.4232 in)
Journal diameter	28.285 $\pm$ 0.05 mm (1.1135 $\pm$ 0.002 in)
Wear limit	28.13 mm (1.107 in)
Camshaft bearing ID	24.97 - 24.98 mm (0.9830 - 0.9835 in)
Camshaft/bearing clearance:	
Standard	0.020 - 0.054 mm (0.0008 - 0.002 in)
Wear limit	0.160 mm (0.006 in)

Crankshaft

Main bearing clearance	0.022 - 0.044 mm (0.0009 - 0.0017 in)
Big-end bearing clearance	0.032 - 0.054 mm (0.0013 - 0.0021 in)
Crankshaft run-out	0.03 mm (0.0012 in)

Clutch

No of plates:	
Plain	6
Inserted	7
No. of springs	6

Inserted plate thickness:									
Standard	...	...	...	...	...	...	...	...	3.0 mm (0.12 in)
Wear limit	...	...	...	...	...	...	...	...	2.8 mm (0.11 in)
Plain plate maximum warpage	...	...	...	...	...	...			0.05 mm (0.002 in)
Spring free length:									
Standard	...	...	...	...	...	...	...	...	42.8 mm (1.685 in)
Service limit	...	...	...	...	...	...	...		41.5 mm (1.634 in)

Gearbox

Type	...	...	...	...	...	...	...	...	5-speed constant mesh
Gear ratios: overall:									
1st	...	...	...	...	...	...	...	...	13.285 : 1
2nd	...	...	...	...	...	...	...	...	8.638 : 1
3rd	...	...	...	...	...	...	...	...	7.069 : 1
4th	...	...	...	...	...	...	...	...	5.955 : 1
5th	...	...	...	...	...	...	...	...	5.201 : 1
Secondary spur gear ratio	...	...	...	...	...	...	...		1.063 : 1 (34/32T)
Middle gear drive ratio	...	...	...	...	...	...			1.056 : 1 (19/18T)

Main torque wrench settings

Cylinder head:									
8 mm nuts	...	...	...	...	...	...	..	...	2.0 kg m (14 ft lbs)
10 mm nuts	...	...	...	...	...	...	...	...	3.5 kg m (25 ft lbs)
Camshaft cover nuts	...	...	...	...	...	...			0.8 - 1.0 kg m (6 - 7 ft lbs)
Crankcase bolts:									
8 mm	...	...	...	...	...	...	...	...	2.0 kg m (14 ft lbs)
10 mm	...	...	...	...	...	...	...	...	3.7 kg m (27 ft lbs)
Big-end cap bolts	...	...	...	...	...	...	...		3.8 kg m (28 ft lbs)
Camshaft cap nuts	...	...	...	...	...	...	...		1.0 kg m (7 ft lbs)

1 General description

The engine fitted to the Yamaha XS 750 models is a double overhead camshaft vertical parallel triple, mounted transversely in the frame. The three throws of the webbed one-piece crankshaft are placed 120° apart, to give an even torque delivery and hence a smooth power output. The big-end bearings are of the split shell type, as are the four main bearings. The left-hand centre main bearing shell in the upper casing incorporates thrust faces to control crankshaft end float.

Integral with the crankshaft are the primary drive chain sprocket, on the extreme right-hand end of the shaft, and the cam chain driven sprocket on the left. The cam chain passes through a tunnel in the cylinder barrel and head, to drive the independent inlet and exhaust camshafts. The cams operate the valves directly through bucket-type followers, into which are fitted the valve clearance adjustment pads.

All engine casings are manufactured in matt black finished aluminium alloy, the cylinder block being fitted with dry cast iron liners.

The crankcases, which house the crankshaft and main gearbox components, separate in a horizontal plane to facilitate dismantling and reassembly.

Wet sump lubrication is supplied by a gear driven trochoid oil pump housed in a detachable casting. The casting also incorporates a wire gauze sediment trap and the oil feed and oil pressure relief valves.

In addition to the gauze screen, fitted to prevent foreign matter from entering the oil pump, the engine is protected by a full-flow oil filter contained within a detachable housing forward of the sump. A by-pass valve integral with the housing retaining bolt prevents oil starvation in the event of filter blockage.

Engine power from the crankshaft is transmitted to the gearbox via an inverted tooth four row chain and a multi-plate clutch. A cone plate shock absorber is mounted outboard of the clutch, to damp out snatch loads in the transmission.

The five-speed constant mesh gearbox is placed to the rear of the crankshaft in the normal manner, but with the mainshaft fitted above the layshaft to reduce overall engine length. A third shaft, incorporating a face cam shock absorber unit, transmits the output to the middle gear casing. This houses the bevel drive gears which turn the drive line through 90° and so enables

the final drive shaft to be interconnected.

The middle gear case is a separate assembly which can be detached from the crankcase as a complete unit.

Electrical power is supplied by a 280 watt alternator, mounted on the right-hand end of the crankshaft, which incorporates an induced magnet rotor excited by a static field coil. Ignition timing is controlled by a set of three contact breakers, each one of which supplies a different cylinder.

2 Operations with engine/gearbox in frame

It is not necessary to remove the engine unit from the frame in order to dismantle the following items:
1 Right and left crankcase covers.
2 Clutch assembly and gear selector components (external).
3 Oil pump and filter.
4 Alternator and starter motor.
5 Cylinder head and cylinder head cover.
6 Cylinder block, pistons and rings.
7 Contact breaker units and starter motor clutch.
8 Middle gear casing assembly.
9 Kickstart assembly, idler gear and engagement ratchet.

3 Operations with engine/gearbox unit removed from frame

As previously described the crankshaft and gearbox assemblies are housed within a common casing. Any work carried out on either of these two major assemblies will necessitate removal of the engine from the frame so that the crankcases can be separated.

4 Removing the engine/gearbox unit

1 Place the machine on its centre stand making sure that it is standing firmly. Although by no means essential it is useful to raise the machine a number of feet above floor level by placing it on a long bench or horizontal ramp. This will enable most of the work to be carried out in an upright position, which is eminently more comfortable than crouching or kneeling in a puddle of sump oil.

2   Place a suitable receptacle below the crankcase and drain off the engine oil. The sump plug lies just to the rear of the oil filter housing. The oil will drain at a higher rate if the engine has been warmed up previously, thereby heating and thinning the oil. Approximately 5½ pints (3 litres) should drain out. Undo the oil filter chamber bolt and remove the chamber and filter, noting that the chamber holds approximately 1 pint. If the middle gear case is to be removed this too should be drained. The drain plug is in the casing rear wall.

3   Raise the dualseat so that it is supported by the bar provided. Dislodge the battery retaining strap and disconnect the negative (Black) lead, followed by the positive (Red) lead. The battery need not be removed unless it is expected that the machine is to be unused for an extended length of time. If this is the case, the battery should be stored safely and given an external charge every two weeks. Note the breather pipe which is a push fit on the union at the left of the battery.

4   Detach the petrol feed pipe and vacuum pipe from the unions at each petrol tap. The tap levers must be placed away from the 'Prime' position as this allows free flow of petrol. Each pipe is secured by a spring clip, the ears of which should be pinched together to release the hold on the pipe. The petrol tank is retained at the rear by a single bolt and cup washer. Remove the bolt and pull the tank rearwards until the forward mounted cups clear the mounting rubbers fitted either side of the frame top tube. Drainage of the tank is not strictly necessary though the reduction in weight may facilitate removal.

5   Release the tachometer drive cable at the front of the cylinder head by unscrewing the knurled ring. The tachometer drive shaft should be removed also, to prevent accidental damage when the engine is being removed. After unscrewing the single socket screw, the retaining plate and drive assembly may be withdrawn.

6   On 750 - 2D models the exhaust system should be removed in the following sequence. Slacken the socket screw which secures the right-hand exhaust pipe to silencer joint. Move the clamp forward along the pipe until it is clear of the joint. Slacken evenly the two right-hand exhaust port flange nuts and slide the flange off the studs. Prise out the two half collars, noting the piece of tape holding them together. The tape was so placed on assembly to hold the collar halves together. Pull the pipe forward away from the port and the silencer joint. Remove the left-hand exhaust port flange and split collars. Slacken the clamp securing the balance pipe joint running below the gearbox. The silencer is retained by two bolts passing through a bracket into the silencer. Remove the bolts and then pull the complete left-hand pipe/silencer unit away from the machine, so that the balance pipes separate. Lift the complete unit away forward to clear the machine. Repeat the process for the central exhaust pipe/silencer assembly. On 750D machines, which are fitted with a three-into-one exhaust system, the removal procedure must be modified slightly. After detaching the mounting bolts and flanges, the system may be lifted away as a complete unit.

7   Remove the frame side covers. Each is a push fit at the base, retained by hooks at the upper edge. Unscrew the single screw holding each intake silencer to the base of the air filter box and pull the silencers away. In order to gain enough clearance to remove the carburettors, the air cleaner box must be moved backwards as far as possible, after unscrewing the four mounting bolts. Note the earth wire is secured by the rearmost left-hand bolt. Loosen the three clamps holding the air hoses to the carburettor mouths and the three clamps which secure the carburettors in the flexible inlet stubs. Pull the cleaner box rearwards and then pull the three carburettors from their stubs. Remove the carburettors towards the left-hand side of the machine, disconnecting the throttle cable from the throttle arm as soon as the necessary access is gained. Detach the breather hose from the air box stub after slackening the screw clips.

8   After removing the carburettors lift the air cleaner box from place, to improve working space. Disconnect the clutch cable at the handlebar control lever after displacing the protective rubber shroud. Pull back the rubber cover on the left-hand engine cover and disconnect the lower end of the cable from the operating arm anchor. Withdraw the cable from the casing.

9   Detach the plastic starter motor cover positioned to the rear of the cylinder block, and disconnect the starter lead, which is held by a nut and washer. Remove also the main earth wire from the rear of the gearbox casing. Disconnect the following block connectors, which are positioned behind the left-hand side cover: Ignition (orange, yellow, grey and blue wires), alternator field coil (green and black wires) and alternator stator (white wires). Pull the wires through so that they are positioned on the engine and will not snag during engine removal. Detach the HT cables from the spark plugs.

10 Carefully prise the drive shaft rubber gaiter off the flange at the rear of the middle gear case, and push it rearwards, so that access may be made to the flanged coupling. Loosen evenly and remove the four flange bolts, turning the flange as required by rotating the rear wheel.

11 Remove the air scoop from the cylinder head, where it is retained on three brackets by six socket screws. Removal of the brackets will give greater (and very necessary) clearance when lifting out the engine. Detach the rear brake pedal from the splined operating shaft after removing the pinch bolt. Do not remove the kickstart lever at this stage as it makes an ideal lifting handle.

12 The engine is supported on three mounting bolts, two short ones at the front and a long one which also secures the rider's footrests, at the rear. When removing the bolts, the engine should be lifted, to prevent damage to the threads. The engine unit is very heavy, and being in close proximity to the adjacent frame tubes not easily removed. At least two people will be required to lift the engine from position if the risk of damage to engine or operator is to be avoided. Manoeuvreing the engine from place is awkward and is likely to be more approximate than precise, and for this reason it is recommended that the lower cradle tubes and the frame tubes adjacent to the front camshaft housing be protected with masking tape. A few moments spent here will render the touching-up of damaged paintwork unnecessary after subsequent reinstallation of the engine. Lift the engine out towards the right-hand side of the machine, taking care that the three main wiring leads do not become snagged, crushed or severed.

4.5 Remove tachometer driveshaft to give clearance

4.6a Detach pipe flange at exhaust port and ...

4.6b ... prise out the half-collars

4.6c Loosen the balance pipe clamp

4.7a Note earth lead on air box retaining bolt

4.7b Detach throttle cable from arm

4.8 Peel back clutch protector and disconnect cable

4.9a Disconnect the main earth lead and also ...

4.9b ... the starter motor cable

4.10 Remove output flange/drive shaft bolts

4.11 Remove the air scoop brackets

4.12a Engine is retained by long through bolt at rear and ...

4.12b ... two short bolts at the front

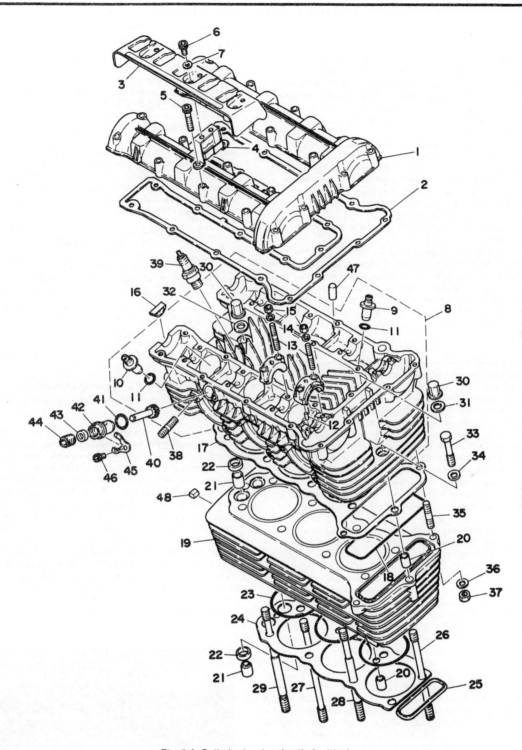

Fig. 1.1. Cylinder head and cylinder block

1	Camshaft cover	13	Stud - 16 off	25	Tunnel seal	37 Nut - 2 off
2	Gasket	14	Plain washer	26	Stud - 3 off	38 Stud - 6 off
3	Air scoop	15	Nut - 16 off	27	Stud	39 Spark plug - 3 off
4	Air scoop bracket - 3 off	16	Plug - 2 off	28	Bolt - 2 off	40 Tachometer drive shaft
5	Socket bolt - 21 off	17	Cylinder head gasket	29	Bolt - 2 off	41 'O' ring
6	Socket bolt - 6 off	18	Tunnel seal	30	Nut - 8 off	42 Housing bush
7	Plain washer - 6 off	19	Cylinder block	31	Plain washer - 8 off	43 Oil seal
8	Cylinder head	20	Hollow dowel - 4 off	32	Heavy plate washer - 2 off	44 Gland nut
9	Inlet valve guide - 3 off	21	Hollow dowel - 4 off	33	Bolt - 2 off	45 Retention claw
10	Exhaust valve guide - 3 off	22	Seal - 4 off	34	Plain washer - 2 off	46 Socket screw
11	'O' ring - 6 off	23	'O' ring - 3 off	35	Stud - 2 off	47 End plug
12	Dowel pin - 4 off	24	Cylinder base gasket	36	Plain washer - 2 off	48 Anti vibration block - 29 off

5 Dismantling the engine/gearbox unit: general

1 Before commencing work on the engine unit, the external surfaces should be cleaned thoroughly. A motorcycle has very little protection from road grit and other foreign matter which sooner or later will find its way into the dismantled engine if this simple precaution is not carried out.

2 One of the proprietary cleaning compounds such as Gunk or Jizer can be used to good effect, especially if the compound is first allowed to penetrate the film of grease and oil before it is washed away. In the USA Gumout degreaser is an alternative.

3 It is essential when washing down to make sure that water does not enter the carburettors or the electrics particularly now that these parts are more vulnerable.

4 Collect together an adequate set of tools in addition to those of the tool roll carried under the seat.

5 Avoid force in any of the operations. There is generally a good reason why an item is difficult to remove, probably due to the use of the wrong procedure or sequence of operations.

6 Dismantling will be made easier if a simple engine stand is constructed that will correspond with the engine mounting points. This arrangement will permit the complete unit to be clamped rigidly to the workbench, leaving both hands free for dismantling.

6 Dismantling the engine/gearbox: removing the camshaft cover and cams

1 Place the engine on the workbench with a small wooden block under the front of the crankcase, to prevent the unit falling forwards. Salcken evenly and remove the camshaft cover socket screws, at the same time detaching the air scoop brackets, if not already removed. If the camshaft cover is stuck to the gasket, great care should be exercised when separating the cover from the cylinder head as it will flex and may fracture. Two lugs are provided to the front and rear of the cover, against which a rawhide mallet may be struck.

2 Unscrew the spark plugs and detach the contact breaker cover from the left-hand side of the engine, where it is retained by three screws. By means of the large hexagon provided, on the end of the contact breaker cam, rotate the engine forwards until the No. 1 (left-hand) cylinder is at TDC with both valves closed. To find TDC view the automatic advance unit through the inspection aperture in the contact breaker base plate. When the T mark to the right of the 1 F mark aligns with the fixed pointer, the correct position has been reached. Note at this stage that the small punch mark on the inner boss of each camshaft sprocket aligns with the projection on the adjacent camshaft bearing cap, indicating that valve timing is correct.

3 Detach the cam chain tensioner as a complete unit, after removing the two retaining nuts. The additional slack now in the chain will allow the upper chain guide to be removed. In order to remove the camshafts, the cam chain must be separated by displacing one of the link pins. This operation should be carried out using a suitable chain-breaker of the type stocked by most large motorcycle dealers. Removal of the pin using a punch and hammer is not recommended as damage to the chain or nearby engine components may easily result. If a top end overhaul only is envisaged, it is imperative that the two ends of the chain are not allowed to fall down through the cam chain tunnel, because they are very difficult to retrieve. Use two lengths of stiff wire to secure the chain ends before pin removal. For the same reason pad the area around the chain and sprockets with clean rag, to prevent the freed pin from disappearing. After removal of the first pin, displace the second pin in the same link to free the link plates. The link **must** be discarded and a new one fitted on reassembly. **Never** re-use a link as the damage incurred due to a chain broken in service will be extreme.

4 Each camshaft in its present position has one lobe opening one valve. To reduce the upward forces and prevent uneven stresses when the camshaft cups are removed, both camshafts must be rotated slightly. Using a spanner applied to the hexagon provided in the middle of each camshaft, turn the exhaust camshaft clockwise exactly 1/6 of a turn and the inlet camshaft anti-clockwise 1/6 of a turn, viewed from the left. Both camshafts will find their own level. Do not rotate either camshaft further than this as the No. 1 piston which is at TDC may be damaged by an opening valve. Likewise **do not rotate the engine.**

5 The camshafts complete with sprockets may be lifted from position after removing the bearing caps, each of which is retained by two nuts. Slacken the cap nuts on each camshaft as evenly as possible from right to left. The caps are marked by letter I for inlet and E for exhaust and numbered 1 - 4, to aid replacement in their original positions.

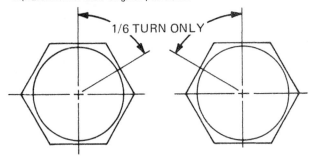

1/6 TURN ONLY

Fig. 1.2. Rotation of camshafts

Rotate camshaft ONLY as shown, prior to bearing cap removal

7 Dismantling the engine/gearbox: removing the cylinder head

1 Remove the camshaft oil feed pipe from the rear of the cylinders by removing the two banjo bolts. Note the position of the copper sealing washers. The cylinder head is a complex structure, and in order to prevent warpage the cylinder head retaining nuts and bolts should be slackened in the correct sequence. Commence by loosening the ten nuts and two bolts ½ a turn each in the sequence shown in the accompanying diagram. Remove all the nuts in sequence and then finally the two bolts. Separation of the cylinder head from the cylinder block may be aided by means of a rawhide mallet. Strike only those portions of the casing which are amply strengthened by webs and lugs. When lifting the cylinder head the two ends of the cam chain should be restrained by long lengths of wire, until the chain can again be secured.

6.2 Align valve timing marks on both camshafts

6.3a Detach the cam chain tensioner

6.3b Press out cam chain link pins

6.3c Use wire to secure chain ends

6.5 Detach bearing caps and lift out both camshafts

7.1a Camshaft oil feed pipe is secured by banjo bolts

7.1b Do not omit middle bolt below RH spark plug or ...

7.1c ... two nuts below cylinder head on left

7.1d Lift head upwards off holding down studs

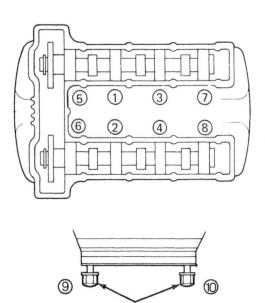

CYLINDER HOLDING NUTS

Fig. 1.3. Cylinder head bolt loosening sequence

confines of the cylinder bores. Endeavour to lift the cylinder block squarely, so that the pistons do not bind in the bores.
3   Remove the two large hollow dowels and seals, noting that the seals are fitted in the same way as the upper seals.
4   Before removing the pistons, mark each on the inside of the skirt to aid identification. It is important that the pistons are refitted to their original cylinders on reassembly. An arrow mark on each piston crown indicates the front, the pistons must therefore be fitted with the arrow facing forwards. Remove the outer circlip from one of the outermost pistons and push out the gudgeon pin. Lift the piston off the connecting rod. The gudgeon pins are a light push fit in piston bosses so they can be removed with ease. If any difficulty is encountered, apply to the offending piston crown a rag over which boiling water has just been poured. This will give the necessary temporary expansion to the piston bosses to allow the gudgeon pin to be pushed out.
5   Each piston is fitted with two compression rings and an oil control ring. It is wise to leave the rings in place on the pistons until the time comes for their examination or renewal in order to avoid confusing their correct order.
6   Remove the two remaining pistons, using a similar procedure.

8   Dismantling the engine/gearbox: removing the cylinder block and pistons

1   Lift the two large hollow dowels from the end of the cylinder barrel top mating surface, noting that the seal on each is fitted with the rounded face downwards. Separate the cylinder block from the base gasket, using the rawhide mallet. If required, a screwdriver may be inserted into the two slots provided in the front of the barrel, to separate the two components. **DO NOT** use screwdrivers or other levers between the mating surfaces; this will certainly lead to oil leaks.
2   Lift the cylinder block upwards off the pistons. At this juncture a second person should be present to support each piston as it leaves the cylinder barrel spigot. To prevent broken particles of piston ring dropping in, and subsequently other foreign matter, all three crankcase mouths should be padded with clean rag. This must be done before the rings leave the

8.1a Remove the hollow dowels and seals

8.1b Use leverage slots to displace cylinder block

8.1c Lift the cylinder block up off pistons

8.4a Prise out and DISCARD circlips

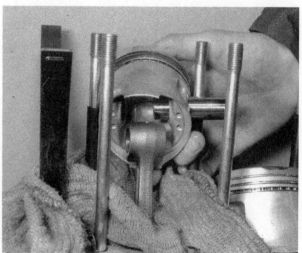

8.4b Push out gudgeon pin to free the piston

9 Dismantling the engine/gearbox: removing the starter motor, contact breaker units and ATU

1 Loosen and remove the two bolts which pass through the starter motor end cap flange into the crankcase. Pull the starter motor across towards the right as far as possible, so that the starter motor boss leaves the aperture in the left-hand casing wall. Raise the motor at the rear so that the starter gear pinion clears the casing and the complete unit can be lifted away. Some difficulty may be encountered when removing the starter motor because the clearance between the end cap and the rear of the primary drive casing is limited. Do not force the motor from position as there is a danger of the top of the motor boss fracturing the left-hand engine casting.

2 The contact breaker baseplate upon which the three sets of points are supported is retained by three screws at the periphery, which pass through elongated holes provided to allow alteration of the ignition timing on one cylinder. Before continuing, mark the relative position of the baseplate and the casing, to aid reassembly. Adjacent punch marks give a clearly read reference. Hold the crankshaft by means of the hexagon provided and loosen the contact breaker cam centre bolt. Remove the bolt

and the hexagon. Detach the neutral indicator switch lead retained at the switch terminal by a single screw, and slacken the lower casing screws which hold the low tension lead clamps. Remove the three contact breaker baseplate screws and lift the complete assembly away.

3 Pull the automatic timing unit (ATU) from place in the end of the crankshaft. The ATU is located in the correct position and driven by a small pin which is a push fit in the end of the crankshaft. Remove the pin and store it in a safe place.

10 Dismantling the engine/gearbox: removing the left-hand engine cover, starter motor driven and oil pump drive gears

1 Loosen and remove the socket screws which retain the left-hand side cover in place. To separate the cover from the crankcase, the use of a rawhide mallet may be required. Pull the cover from place to give access to the drive gears. The starter intermediate gear and spindle may come away either in the cover or may remain in the gear chest. In either event pull the spindle and gear from position.

2 The oil pump drive gear, which also serves as the starter motor final driven gear, is secured to the crankshaft by a large

nut and washer. Lock the crankshaft by passing a close fitting bar through the small end eye of one connecting rod so that the bar bears down on two small wooden blocks placed across the relevant crankcase mouth. Remove the nut - which is very tight - and draw the gear off the shaft. Note the five pins projecting from the rear of the pinion, which engage with the sprocket form on the crankshaft.

3   Pull the starter clutch, double gear and oil pump driveshaft from the casing as a complete unit. The starter clutch, which forms the main body of this assembly, may be dismantled at a later stage, if required.

4   Remove the oil delivery pipe from within the casing after unscrewing the banjo bolt passing through each union.

11 Dismantling the engine/gearbox: removing the cam chain and chain guides

1   The cam chain may now be removed by rotating the crankshaft slowly and feeding the chain off the sprocket. Each chain guide is supported at the lower end by a single screw. Remove both guides, noting the pivot bush and sequence of washers on each.

9.1 Lift starter motor out very carefully

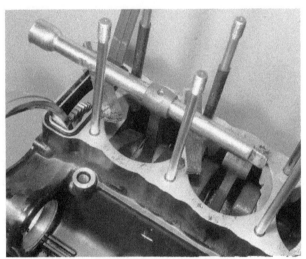

10.2a Lock the crankshaft using a long bar and ...

10.2b ... loosen the oil pump drive pinion nut

10.3 Remove oil pump shaft/starter clutch as a unit

12 Dismantling the engine/gearbox: removing the alternator

1   Tip the engine forwards so that access is available to the underside of the crankcase. Loosen the three main wiring lead clamps and free the leads. Disconnect the auxilliary lead to the oil pressure switch either by separating the snap connector or by removing the terminal screw from the switch.

2   Remove the alternator cover, which is retained by six socket screws. The cover contains the alternator field coils and the main stator coils.

3   Lock the crankshaft, using the bar through the small end eye method, and remove the alternator centre bolt and washers. The alternator rotor is a tight tapered fit on the end of the crankshaft and will therefore need drawing from position by using a suitable extractor. The rotor is threaded to accept a special bolt which, when tightened, will push the complete unit from position. If a metric bolt of equivalent size and thread pitch is to hand, this may be used in place of the special service tool. To protect the crankshaft end and the internal thread into which the alternator centre bolt normally screws, a plug or button should be inserted before fitting and tightening the extractor bolt. The plug can be fashioned from a suitable small

bolt. A standard two or three legged sprocket puller may also be used to remove the rotor. In this instance insert the centre bolt before fitting the puller. The bolt should be screwed in lightly. Under no circumstances should levers be used in an attempt to remove the rotor. Such an approach will almost certainly lead to damage of the rotor or to the adjacent casings.

13 Dismantling the engine/gearbox: removing the primary drive, clutch and kickstart mechanism

1   Detach the kickstart from the splined shaft after removing completely the pinchbolt. Unscrew the primary drive cover retaining screws and lift the cover away. A rawhide mallet may be required to separate the cover initially.
2   An outrigger bearing and cast support bracket are fitted to the outer end of the gearbox mainshaft. Remove the securing screws and pull the bracket, complete with bearing, away from the shaft and case. Fitted below the outrigger, outboard of the clutch, is a shock absorber unit comprising a face cam within a drum, under tension from cone shaped plate. The complete unit may be prised from the centre of the primary driven sprocket after removal of the small circlip and washer followed by the large circlip. All these components are fitted to the end of the gearbox mainshaft. Use two small screwdrivers between the circlip fitted to the drum periphery and the sprocket, to lever the unit from position.
3   The primary driven sprocket, together with the drive chain, are now free to be lifted away. Pull the clutch outer drum off the shaft to gain access to the clutch main assembly. Remove the thrust washer and circlip from the mainshaft adjacent to the clutch pressure plate.
4   Unscrew the clutch spring bolts evenly and remove them, together with the pressure springs. Lift the clutch pressure plate from place and then remove the clutch plates, either one at a time or as an assembly. Note the sequence of plates, to aid reassembly. The innermost plain plate is secured by a circlip which need not be disturbed at this juncture. Displace the washer and clutch operating crossbar from the mainshaft. Push the clutch pushrod in from the left-hand side of the engine so that the small ball bearing is displaced from the hollow main-shaft. Withdraw the pushrod from the left.
5   Bend down the ears of the tab washer which secures the clutch centre boss nut. Because of the length of mainshaft protruding, an ordinary socket will not fit the nut. A special long socket is available as a special service tool, but if this cannot be obtained, a 32 mm box spanner will serve equally well. To prevent the clutch centre boss rotating when loosening the nut, a locking sprag may be fabricated from a piece of steel sheet and placed between the casing and one of the splines on the clutch boss. See the accompanying photograph. Care should be taken that the sprag is placed securely against a well supported portion of the casing. Alternatively, place the gearbox in top gear and hold the output shaft flange. This may be accomplished with ease by inserting two bolts in diametrically opposed holes in the flange and placing a long bar across the bolt heads.
6   With the nut removed, withdraw the tab washer, conical spring washer, plain washer (where fitted) and the clutch boss, followed by the spacer and washer.
7   Remove the large circlip which retains the kickstart idler gear to the stub end of the middle gear intermediate reduction shaft. The circlip is backed by a large shim. Using a screwdriver, disengage the outer turned end of the tensioned kickstart spring from the anchor lug in the casing. The kickstart shaft assembly should be removed from the casing simultaneously with the kickstart idler gear.

14 Dismantling the engine/gearbox: removing the sump and oil pump

1   Tip the engine forwards so that access is available to the underside of the engine. Loosen evenly and remove the sump retaining screws followed by the sump itself. The oil pump is incorporated in a casting which also contains a filter screen and two one-way plunger valves. It is retained by three socket screws. On all but early models one screw is obscured by the filter screen mouth, which must be detached to gain access. Remove the screen, which itself is held by three screws and then detach the bell shaped casting, which is held by two screws. After removal of the three main screws, lift the oil pump away complete.

15 Dismantling the engine/gearbox: removing the middle gear unit

1   The middle gearbox which houses the 90° bevel drive gears is detachable from the crankcase as a complete unit, and is retained by seven socket bolts of varying lengths. The bolts should be slackened evenly and then removed. Two different types of gearbox are fitted, depending on the country of original delivery and availability. The first type has three socket bolts passing through the outer bearing end cap and the second type is fitted with four bolts. On the three bolt cap the lower bolt serves to retain the gearbox and on the four bolt cap the forward of the two bolts is also a gearbox retaining bolt.
2   As the seven bolts are unscrewed the middle gear case will be pushed away from the crankshaft by the cam type shock absorber spring, which is fitted to the middle gear input shaft splines. After removal of the casing, the cam and spring can be pulled off the shaft.

16 Dismantling the engine/gearbox: separating the crankcase halves

1   The two crankcase halves are retained together by bolts passing through from both the upper and lower cases. The upper bolts are numbered 24 - 15 and the bolts in the lower case 14 - 1. In addition there is one extra bolt that **MUST NOT** be over-looked. This bolt is located in the primary drive casing, between the gearbox mainshaft and the middle gear intermediate shaft. To avoid the risk of distortion, the bolts should be loosened in sequence, starting with the unnumbered bolt followed by the remainder of the bolts, highest numbers first. Slacken each bolt ½ a turn initially, and then loosen the bolts fully, still in sequence. Note that bolts numbered 5, 6, 7 and 8 are fitted with copper washers.
2   Having removed all the bolts, place the crankcase so that the base is resting on the workbench. Use a rawhide mallet to separate the casing halves and then lift the upper half away, leaving the crankshaft and gearbox components resting in the lower case. Be prepared to support each connecting rod as it is cleared by the upper casing. Endeavour to prevent the main bearing shells in the upper casing from dropping out. If the shells become displaced, refit them immediately in their original positions.

17 Dismantling the engine/gearbox: removing the crankshaft

1   Removal of the crankshaft is quite straightforward. Lift the crankshaft upwards out of the bearing lower shells, leaving the shells behind in the casing. Displaced shells should be refitted immediately in their original positions, until the examination stage.

18 Dismantling the engine/gearbox: removing the gear shaft assemblies and selector mechanism

1   Lift the middle gear input shaft from the casing complete with the large oil seal and the two bearings. Note the half clips which locate the bearings in the casing in an axial plane. Lift the clips from position.

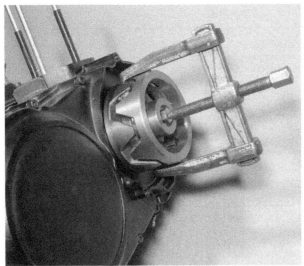

12.3 Use sprocket puller to remove alternator rotor

13.2a Clutch outrigger held by four bolts

13.2b Pull shock absorber from place and ...

13.3... lift off primary chain and driven sprocket

13.4 Remove heavy washer and cross-bar

13.5 Use steel sprag when loosening clutch nut

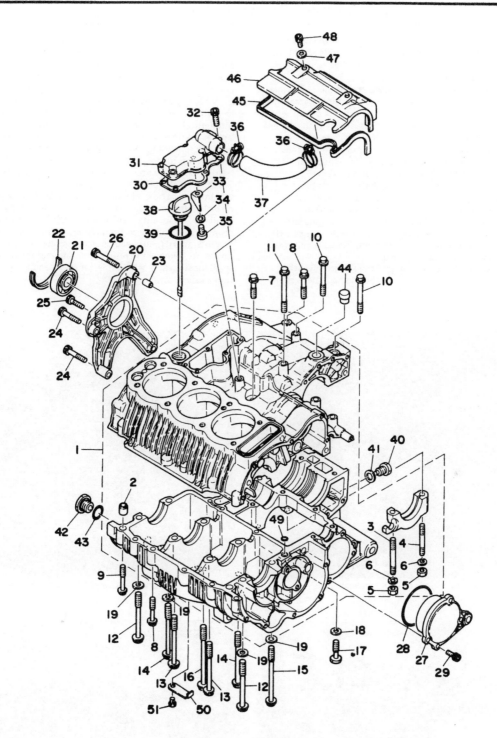

Fig. 1.4. Crankcase components

1 Crankcase assembly	13 Bolt - 2 off	27 Bearing housing	39 'O' ring
2 Hollow dowel - 2 off	14 Bolt - 2 off	28 'O' ring	40 Drain plug
3 Stud	15 Bolt	29 Socket screw - 4 off	41 Sealing washer
4 Stud	16 Bolt	30 Gasket	42 Oil pressure switch adaptor
5 Nut - 2 off	19 Plain washer - 4 off	31 Breather cover	43 'O' ring
6 Plain washer	20 Outrigger bearing support	32 Socket screw - 6 off	44 Breather
7 Bolt	bracket	33 Breather plate	45 Seal strip
8 Bolt - 11 off	21 Journal ball bearing	34 Spring washer	46 Starter motor cover
9 Bolt	23 Hollow dowel - 2 off	35 Screw	47 Plain washer - 2 off
10 Bolt - 2 off	24 Socket bolt - 2 off	36 Hose clip - 2 off	48 Socket screw - 2 off
11 Bolt	25 Socket bolt	37 Breather hose	49 'O' ring
12 Bolt - 2 off	26 Socket bolt	38 Oil level dipstick /filler	50 Cable clip - 3 off
		plug	51 Bolt - 3 off

2   Displace the E clip from the right-hand end of the upper selector fork rod. Withdraw the rod towards the left and lift out the single selector fork. The mainshaft is now free to be lifted from position as a complete assembly.

3   Prise off the E clip from the left-hand end of the gearchange shaft, where it protrudes from the casing. Grasp the gearchange shaft at the quadrant end and withdraw it from the casing, complete with the centraliser spring. The quadrant with which the gearshaft meshes is retained on its pivot shaft by an E clip. Remove the clip and hold back the selector pawl to clear the pins in the change drum. Pull the quadrant from position. Removal of the quadrant pivot shaft is not necessary.

4   Invert the crankcase so that the layshaft is accessible. Prise the E clip from the selector fork rod and pull the rod out towards the left, to free the two selector forks. Note the washer which is fitted between the bearing cap rod support lug and the E clip. Each selector fork has a push fit pin which engages with the appropriate channel in the change drum. Remove and store the pins safely; they are easily mislaid.

5   Remove the four screws which hold the flanged layshaft bearing housing to the crankcase. The housing has a deep spigot and is a tight push fit; care must be taken when easing it from position. Unscrew the socket screw and remove the two washers from the opposite end of the layshaft. Use a length of steel plate as a sprag, engaged with a tooth on any suitable gear pinion, to prevent rotation of the shaft whilst loosening the screw.

6   Move the complete layshaft over until the final output gear - located between the centre bearing and gearbox wall - can be lifted out. Pull the shaft out further and slide off the plain washer followed by the 1st gear pinion. The layshaft, complete with the remaining four gear pinions, can now be removed through the tunnel in the gearbox side. Unscrew the centre bearing cap nuts and remove the cap, together with the bearing.

7   Unscrew the change drum locating bolt from its location within the gearbox. The bolt is secured by a tab washer, the ears of which must be bent down before loosening the bolt. Similarly, remove the neutral indicator switch and the change drum detent bolt, spring and plunger. Slide the change drum out towards the left, until access can be made to the drum stopper quadrant. Remove the retaining circlip and the quadrant. The small locating and drive pin is a push fit and is easily lost. Store it in a safe place.

16.2a Do not forget to remove hidden crankcase bolt

16.2b Lift the upper case away leaving components in place

18.1 Lift middle gear input shaft from casing

18.2a Withdraw the upper selector rod and fork

18.2b Lift out the mainshaft as a unit

18.3 Pull out the gearchange arm from case

18.4a Note washer fitted behind rod securing circlip

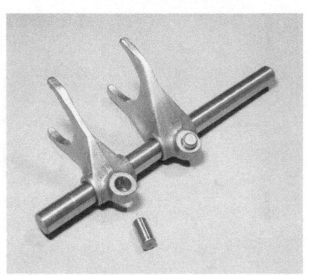
18.4b Do not lose fork guide pins

18.5a Ease layshaft bearing housing from position

18.5b Lock layshaft with lever when loosening end bolt

18.6a Lift out final output gear and then ...

18.6b ... slide off washer and 1st gear pinion

18.7a Remove the change drum locating bolt ...

18.7b ... the neutral indicator switch and ...

18.7c ... the drum detent bolt, spring and plunger

18.7d Remove circlip to detach drum stopper quadrant

19 Examination and renovation: general

1   Before examining the component parts of the dismantled engine/gear unit for wear, it is essential that they should be cleaned thoroughly. Use a paraffin/petrol mix to remove all traces of oil and sludge which may have accumulated within the engine.

2   Examine the crankcase castings for cracks or other signs of damage. If a crack is discovered, it will require professional attention or in an extreme case, renewal of the casting.

3   Examine carefully each part to determine the extent of wear. If in doubt, check with the tolerance figures whenever they are quoted in the text. The following sections will indicate what type of wear can be expected and in many cases, the acceptable limits.

4   Use clean, lint-free rags for cleaning and drying the various components, otherwise there is risk of small particles obstructing the internal oilways.

20 Examination and renovation: main bearings and big-end bearings

1   The Yamaha 750 is fitted with shell type bearings on the crankshaft and the big-end assemblies.

2   Bearing shells are relatively inexpensive and it is prudent to renew the entire set of main bearing shells when the engine is dismantled completely, especially in view of the amount of work which will be necessary at a latter date if any of the bearings fail. Always renew the four (4) sets of main bearings together. Note that the bearing shell second from the left in the upper casing half differs in form from the remainder. This shell, in addition to serving as a bearing in the normal manner, also controls crankshaft end float.

3   Wear is usually evident in the form of scuffing or score marks in the bearing surface. It is not possible to polish these marks out in view of the very soft nature of the bearing surface and the increased clearance that will result. If wear of this nature is detected, the crankshaft must be checked for ovality as described in the following Section.

4   Failure of the big-end bearings is invariably accompanied by a pronounced knock within the crankcase. The knock will become progressively worse and vibration will also be experien-

ced. It is essential that bearing failure is attended to without delay because if the engine is used in this condition there is a risk of breaking a connecting rod or even the crankshaft, causing more extensive damage.

5   Before the big-end bearings can be examined the bearing caps must be removed from each connecting rod. Each cap is retained by two high tensile bolts. Before removal, mark each cap in relation to its connecting rod so that it may be replaced correctly. As with the main bearings, wear will be evident as scuffing or scoring and the bearing shells must be replaced as complete sets.

6   Replacement bearing shells for either the big-end or main bearings are supplied on a selected fit basis (ie; bearings are selected for correct tolerance to fit the original journal diameter), and it is essential that the parts to be used for renewal are of identical size. Code numbers stamped on various components are used to identify the correct replacement bearings for both the crankshaft, main bearing and the big-end journals. The journal size numbers are stamped on the crankshaft left-hand outside web, the first four numbers being the main bearing journal numbers and the following three those of the big-end journal. The main bearing housing numbers are stamped on the front mating surface of the upper crankcase half. To ascertain which main bearing insert is required, subtract the crankcase number for that particular bearing from the crankshaft journal number. The resultant figure should be compared against the following table, in order to select the correct colour code of shell bearing:

Insert	Colour Code
No. 1	Blue
No. 2	Black
No. 3	Brown
No. 4	Green
No. 5	Yellow

A similar method is used to select the correct big-end shell bearings. Each connecting rod is marked in ink on the rod and the cap. Subtract the big-eng journal number from the appropriate connecting rod number, and use the number obtained to select the correct colour coded bearing shell set from the table. In practice, provided the coded numbers are quoted correctly, the Yamaha specialist supplying the replacement parts will make the necessary calculations.

20.2a Standard main bearing shell

20.2b Single thrust main bearing in upper casing half

20.5a Separate the connecting rods from the big-end caps

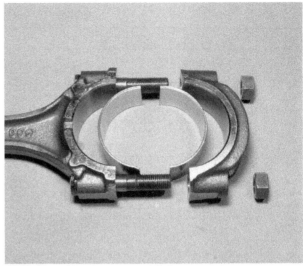

20.5b Big-end bearing set: general view

20.6a Crankshaft journal size numbers on left of web

20.6b Big-end journal numbers on right of web

20.6c Main bearing housing number stamped on upper casing

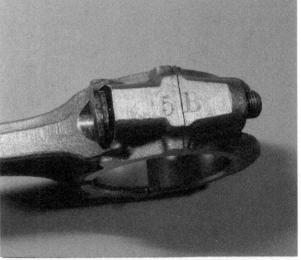

20.6d Big-end bearing housing number etched on connecting rod

21 Examination and renovation: crankshaft assembly

1 If wear has necessitated the renewal of the big-end and/or main bearing shells, the crankshaft should be checked with a micrometer to verify whether ovality has occurred. If the reading on any one journal varies by more than 0.04 mm (0.0015 in) the crankshaft should be renewed.

2 Mount the crankshaft by supporting both ends on V blocks or between centres on a lathe and check the run-out at the centre main bearing surfaces by means of a dial gauge. The run-out will be half that of the gauge reading indicated. A measured run-out of more than 0.03 mm (0.001 in) indicates the need for crankshaft renewal. It is wise, however, before taking such drastic (and expensive) action, to consult a Yamaha specialist.

3 The clearance between any set of bearings and their respective journal may be checked by the use of plastigauge (press gauge). Plastigauge is a graduated strip of plastic material that can be compressed between two mating surfaces. The resulting width of the material when measured with a micrometer will give the amount of clearance. For example if the clearance in the big-end bearing was to be measured, plastigauge should be used in the following manner.

Cut a strip of plastigauge to the width across the bearing to be measured. Place the plastigauge strip across the bearing journal so that it is parallel with the crankshaft. Place the connecting rod complete with its half shell on the journal and then carefully replace the bearing cap complete with half shell onto the connecting rod bolts. Replace and tighten the retaining nuts to the correct torque and then loosen and remove the nuts and the bearing cap. Without bending or pressing the plastigauge strip, place it at its thickest point between a micrometer and read off the measurement. This will indicate the precise clearance. The original size and wear limit of the crankshaft journals and the standard and service limit clearance between all the bearings is given in the Specifications at the beginning of this Chapter. Clearances may be checked also by direct measurement of each journal and bearing using external and internal micrometers.

4 The crankshaft has drilled oil passages which allow oil to be fed under pressure to the working surfaces. Blow the passages out with a high pressure air line to ensure they are absolutely free. Blanking plugs in the form of small steel balls are fitted in each web, to close off the outer ends of the passages. Check that these balls, which are peened into place, are not loose. A plug coming free in service will cause oil pressure loss and resultant bearing and journal damage.

5 When refitting the connecting rods and shell bearings, note that under no circumstances should the shells be adjusted with a shim 'scraped in' or the fit 'corrected' by filing the connecting rod and bearing cap or by applying emery cloth to the bearing surface. Treatment such as this will end in disaster; if the bearing fit is not good, the parts concerned have not been assembled correctly. This advice also applies to the main bearing shells. Use new big-end bolts too - the originals may have stretched and weakened.

6 Oil the bearing surfaces before reassembly takes place and make sure the tags of the bearing shells are located correctly. After the initial tightening of the connecting rod nuts, check that each connecting rod revolves freely, then tighten to a torque setting of 3.8 kg m (27 ft lbs). Check again for ease of rotation.

22 Examination and renewal: oil seals

1 An oil seal is fitted to the primary drive cover to prevent oil loss into the alternator chamber. Similarly a seal is fitted to the starter gear chest, to protect the contact breakers. The middle gear input shaft is fitted with a special double lipped seal which prevents oil from the engine from contaminating the Hypoy gear oil in the middle gearbox. In addition, the kickstart shaft, gear selector shaft and clutch pushrod are also fitted with seals. Check all the seals for damage to the feathered lip and the

21.5 Torque the big-end bolts to the specified setting

garter spring. If wear is evident, or leakage has been experienced, the seal in question should be renewed.

2 Oil seals also tend to lose their effectiveness if they harden with age. It is difficult to give any firm recommendations in this respect except to say that if there is any doubt about the condition of a seal, renew it as a precaution.

3 Oil seals fitted within a housing in a casing are a light drive fit and may be drifted from position, using a tubular drift of suitable outside diameter. When fitting a new seal, ensure that it is driven in squarely. As a general rule the spring garter side of the seal should face towards the fluid to be contained.

23 Examination and renovation: connecting rods

1 It is unlikely that any of the connecting rods will become damaged during normal usage unless an unusual occurrence such as a dropped valve causes the engine to lock. This may well bend the connecting rod in that cylinder. Carelessness when removing a tight gudgeon pin can also give rise to a similar problem. It is not advisable to straighten a bent connecting rod; renewal is the only satisfactory solution.

2 The bearing surface of each small end eye is provided by a cold-metal-sprayed coating with a bronze base. If the small end eye wears, the connecting rod in question must be renewed. If the clearance between a gudgeon pin and small end is excessive, check first that the wear is in the eye and not the gudgeon pin. This will prevent the unnecessary renewal of a sound component. Always check that the oil hole in the small end eye is not blocked since if the oil supply is cut off, the bearing surfaces will wear very rapidly.

24 Examination and renovation: cylinder bores

1 The usual indication of badly worn cylinder bores and pistons is excessive smoking from the exhausts, high crankcase compression which causes oil leaks, and piston slap, a metallic rattle that occurs when there is little or no load on the engine. If the top of the cylinder bore is examined carefully, it will be found that there is a ridge at the front and back the depth of which will indicate the amount of wear which has taken place. This ridge marks the limit of travel of the top piston ring.

2 Since there is a difference in cylinder wear in different directions, side to side and back to front measurements should be made. Take measurements at three different points down the length of the cylinder bore, starting just below the top piston ring ridge, then about 60 mm (2½ in) below the top of the

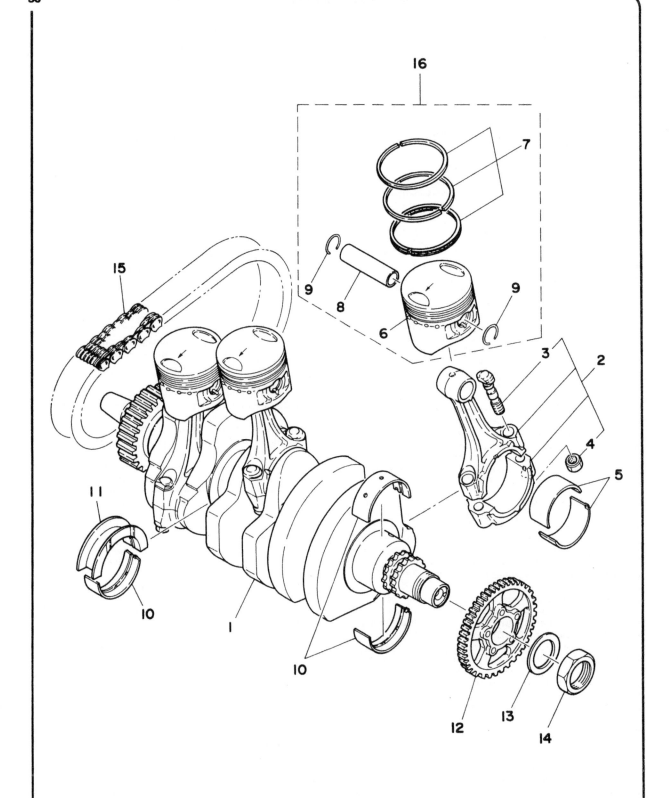

Fig. 1.5. Crankshaft component parts

1	Crankshaft	5	Big-end bearing shell set - 3 off	9	Circlip - 6 off	13	Plain washer
2	Connecting rod assembly - 3 off			10	Main bearing shell - 7 off	14	Nut
		6	Piston - 3 off	11	Main bearing thrust shell	15	Nut
3	Special bolt - 6 off	7	Piston ring set - 3 off	12	Oil pump drive pinion	16	Piston complete - 3 off
4	Nut - 6 off	8	Gudgeon pin - 3 off				

bore and the last measurements about 25 mm (1 in) from the bottom of the cylinder bore. The cylinder measurement as standard and the service limit are as follows:

Standard	Service limit
Cylinder bore	
68.00 - 68.02 mm	68.10 mm
(2.677 - 2.678 in)	(2.681 in)

If any of the cylinder bore inside diameter measurements exceed the service limit the cylinder must be bored out to take the next size of piston. If there is a difference of more than 0.05 mm (0.002 in) between any two measurements the cylinder should, in any case, be rebored.

3   Oversize pistons are available in four oversizes: 0.25 mm (0.009 in); 0.50 mm (0.020 in); 0.75 mm (0.030 in) and 1.0 mm (0.040 in).

4   Check that the surface of the cylinder bore is free from score marks or other damage that may have resulted from an earlier engine seizure or a displaced gudgeon pin. A rebore will be necessary to remove any deep scores, irrespective of the amount of bore wear that has taken place, otherwise a compression leak will occur.

5   Make sure the external cooling fins of the cylinder block are not clogged with oil or road dirt which will prevent the free flow of air and cause the engine to overheat.

6   If removed for any reason, the cylinder block and cylinder head holding down studs are removed from the crankcase, they should be smeared with Loctite before they are reinserted. This is essential if risk of oil seepage is to be avoided, as some of the studs pierce the oil galleries.

25 Examination and renovation: pistons and piston rings

1   Attention to the pistons and piston rings can be overlooked if a rebore is necessary, since new components will be fitted.

2   If a rebore is not necessary, examine each piston carefully. Reject pistons that are scored or badly discoloured as the result of exhaust gases by-passing the rings.

3   Remove all carbon from the piston crowns, using a blunt scraper, which will not damage the surface of the piston. Clean away carbon deposits from the valve cutaways and finish off with metal polish so that a smooth, shining surface is achieved. Carbon will not adhere so readily to a polished surface.

4   Small high spots on the back and front areas of the piston can be carefully eased back with a fine swiss file. Dipping the file in methylated spirits or rubbing its teeth with chalk will prevent the file clogging and eventually scoring the piston. Only very small quantities of material should be removed, and never enough to interfere with the correct tolerances. Never use emery paper or cloth to clean the piston skirt; the fine particles of emery are inclined to embed themselves in the soft aluminium and consequently accelerate the rate of wear between bore and piston.

5   Measure the outside diameter of the piston about 10.0 mm (0.4 in) up from the skirt at right angles to the line of the gudgeon pin. To determine the piston/cylinder barrel clearance, subtract the maximum piston measurement from the minimum bore measurement. If the clearance exceeds 0.1 mm (0.004 in), the piston should ideally be renewed. This however, is seeking perfection, and an additional clearance of perhaps 0.025 mm (0.001 in) will not reduce engine performance dramatically.

6   Check that the gudgeon pin bosses are not worn or the circlip grooves damaged. Check that the piston ring grooves are not enlarged. Side float should not exceed 0.08 mm (0.003 in) for the top ring and 0.07 mm (0.0028 in) for the second ring. Side play on the oil control ring is not measurable.

7   Piston ring wear can be measured by inserting the rings in the bore from the top and pushing them down with the base of the piston so that they are square with the bore and close to the bottom of the bore where the cylinder wear is least. Place a feeler gauge between the ring end. If the clearance exceeds the service limit the ring should be renewed. The expander band of the oil control ring cannot be measured. In practice, if wear of the two side rails exceeds the limit, the three components should be renewed. It is advised that, provided new rings have not been fitted recently, a complete set of rings be fitted as a matter of course whenever the engine is dismantled. This action will ensure maintenance of compression and performance. If new rings are to be fitted to cylinder bores which are in good condition and do not require a rebore, it is essential to have the surface of the bores honed lightly. This operation is known as glazebusting and as the name suggests, it removes the mirror smooth surface which has been produced by the previous innumerable up and down strokes of the piston and rings. If the glaze is not removed, the new rings will glide over the surface, making the running-in process unnecessarily protracted.

8   Check that there is no build up of carbon either in the ring grooves or the inner surfaces of the rings. Any carbon deposits should be carefully scraped away. A short length of old piston ring fitted with a handle and sharpened at one end to a chisel point is ideal for scraping out encrusted piston ring grooves.

9   All pistons have their size stamped on the piston crown, original pistons being stamped standard (STD) and oversize pistons having the amount of oversize indicated. Similarly oversize piston rings are stamped on the upper edge.

26 Examination and renovation: cylinder head and valves

1   Before dismantling the valve gear proper lift out each of the six bucket type cam followers, together with the adjustment pads. Ensure that each follower is marked clearly, so that it may be replaced in the original recess in the cylinder head. Marking of the adjuster pads is not required. It is best to remove all carbon deposits from the combustion chambers before removing the valves for inspection and grinding-in. Use a blunt ended chisel or scraper so that the surfaces are not damaged. Finish off with a metal polish to achieve a smooth, shining surface. If a mirror finish is required, a high speed felt mop and polishing soap may be used. A chuck attached to a flexible drive will facilitate the polishing operation.

2   A valve spring compression tool must be used to compress each set of valve springs in turn, thereby allowing the split collets to be removed from the valve cap and the valve springs and caps to be freed. Keep each set of parts separate and mark each valve so that it can be replaced in the correct combustion chamber. There is no danger of inadvertently replacing an inlet valve in an exhaust position, or vice-versa, as the valve heads are of different sizes. The normal method of marking valves for later identifica-

26.1 Lift out each of the bucket type cam followers

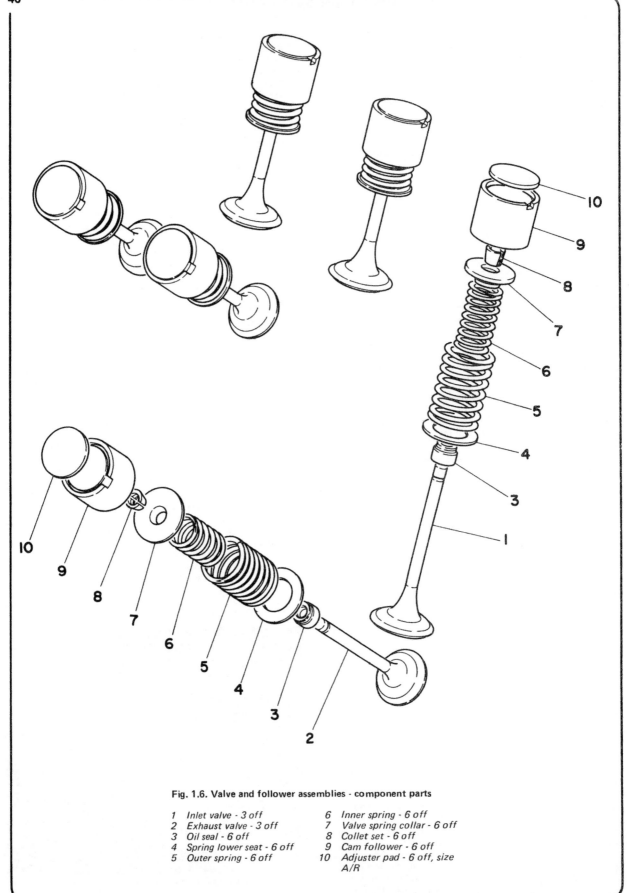

Fig. 1.6. Valve and follower assemblies - component parts

1	Inlet valve - 3 off	6	Inner spring - 6 off
2	Exhaust valve - 3 off	7	Valve spring collar - 6 off
3	Oil seal - 6 off	8	Collet set - 6 off
4	Spring lower seat - 6 off	9	Cam follower - 6 off
5	Outer spring - 6 off	10	Adjuster pad - 6 off, size A/R

tion is by centre punching them on the valve head. This method is not recommended on valves, or any other highly stressed components, as it will produce high stress points and may lead to early failure. Tie-on labels, suitably inscribed, are ideal for the purpose. Because of the cylinder head design, modification of an existing valve spring compressor may be necessary so that it clears the high walls of the cam and valve spring compartments. Remove the oil seal cap from each valve guide.

3 Before giving the valves and valve seats further attention, check the clearance between each valve stem and the guide in which is operates. Clearances are as follows:

Standard	Service limit
Inlet valve/guide clearance	
0.020 - 0.041 mm	0.10 mm (0.004 in)
(0.0008 - 0.0016 in)	
Exhaust valve/guide clearance	
0.035 - 0.059 mm	0.12 mm (0.005 in)
(0.0014 - 0.0023 in)	

Measure the valve stem at the point of greatest wear and then measure again at right-angles to the first measurement. If the valve stem is below the service limit it must be renewed.

Standard (max)	Service limit
Inlet valve stem	
7.010 mm (0.2760 in)	6.975 mm (0.2746 in)
Exhaust valve stem	
7.025 mm (0.2765 in)	6.960 mm (0.2740 in)

The valve stem/guide clearance can be measured with the use of a dial gauge and a new valve. Place the new valve into the guide and measure the amount of shake with the dial gauge tip resting against the top of the stem. If the amount of wear is greater than the wear limit, the guide must be renewed.

4 To remove an old valve guide, place the cylinder head in an oven and heat it to about 100°C (212°F). The old guide can now be tapped out from the cylinder side. The correct drift should be shouldered with the smaller diameter the same size as the valve stem and the larger diameter slightly smaller than the O.D. of the valve guide. If a suitable drift is not available a plain brass drift may be utilised with great care. Even heating is essential, if warpage of the cylinder head is to be avoided. If in doubt, seek the advice of a Yamaha specialist. Each valve guide is fitted with an 'O' ring to ensure perfect sealing. The 'O' rings must be replaced with new components. New guides should be fitted with the head at the same heat as for removal. Restore the correct valve/guide clearance using a 7 mm reamer.

5 Valve grinding is a simple task. Commence by smearing a trace of fine valve grinding compound (carborundum paste) on the valve seat and apply a suction tool to the head of the valve. Oil the valve stem and insert the valve in the guide so that the two surfaces to be ground in make contact with one another. With a semi-rotary motion, grind in the valve head to the seat, using a backward and forward action. Lift the valve occasionally so that the grinding compound is distributed evenly. Repeat the application until an unbroken ring of light grey matt finish is obtained on both valve and seat. This denotes the grinding operation is now complete. Before passing to the next valve, make sure that all traces of the valve grinding compound have been removed from both the valve and its seat and that none has entered the valve guide. If this precaution is not observed, rapid wear will take place due to the highly abrasive nature of the carborundum base.

6 When deep pits are encountered, it will be necessary to use a valve refacing machine and a valve seat cutter, set to an angle of 45°. Never resort to excessive grinding because this will only pocket the valves in the head and lead to reduced engine efficiency. If there is any doubt about the condition of a valve, fit a new one.

7 Examine the condition of the valve collets and the groove on the valve stem in which they seat. If there is any sign of damage, new parts should be fitted. Check that the valve spring collar is not cracked. If the collets work loose or the collar splits

whilst the engine is running, a valve could drop into the cylinder and cause extensive damage.

8 Check the free length of each of the valve springs. The springs should be renewed, preferably as a complete set, if any has compressed to, or beyond, the service limit of 33.6 mm/1.3228 in (inner) or 37.9 mm/1.4921 in (outer).

9 Reassemble the valve and valve springs by reversing the dismantling procedure. Fit new oil seals to each valve guide and oil both the valve stem and the valve guide, prior to reassembly. Take special care to ensure the valve guide oil seal is not damaged when the valve is inserted. As a final check after assembly, give the end of each valve stem a light tap with a hammer, to make sure the split collets have located correctly.

10 Check the cylinder head for straightness, especially if it has shown a tendency to leak oil at the cylinder head joint. If there is any evidence of warpage, provided it is not too great, the cylinder head must be either machined flat or a new head fitted. Most cases of cylinder head warpage can be traced to unequal tensioning of the cylinder head nuts and bolts by tightening them in incorrect sequence or using incorrect or unmeasured torque settings.

27 Examination and renovation: camshafts, camshaft bearings and cam followers

1 The camshaft should be examined visually for wear, which will probably be most evident on the ramps of each cam and where the cam contour changes sharply. Also check the bearing surfaces for obvious wear and scoring. Cam lift can be checked by measuring the height of the cam from the bottom of the base circle to the top of the lobe. If the measurement is less than the service limit which is as follows the opening of that particular valve will be reduced resulting in poor performance.

Standard	Service limit
Inlet	
36.803 ± 0.05 mm	36.75 mm (1.446 in)
(1.448 ± 0.002 in)	
Exhaust	
36.303 ± 0.05 mm	36.15 mm (1.423 in)
(1.429 ± 0.002 in)	

Measure the diameter of each bearing journal with a micrometer or vernier gauge. If the diameter is less than the service limit, renew the camshaft.

2 The camshaft bears directly on the cylinder head material and that of the bearing caps, there being no separate bearings. Check the bearing surfaces for wear and scoring. The clearance between the camshaft bearing journals and the aluminium bearing surfaces may be checked using plastigauge (press gauge) material in the same manner as described for crankshaft bearing clearance in Section 20.3 of this Chapter. If the clearance is greater than given for the service limit the recommended course is to replace the camshaft. If bad scuffing, is evident on the camshaft bearing surfaces, due to a lubrication failure, the only remedy is to renew the cylinder head and bearing caps and the camshaft if it transpires that it has been damaged also.

3 Inspect the outer surface of the cam followers for evidence of scoring or other damage. If a cam follower is in poor condition, it is probable that the guideway in which it works is also damaged. In extreme cases this may necessitate renewal of both the follower and the cylinder head. Check for clearance between the followers and their guideways. If excessive slack is evident, renew the follower.

28 Examination and renovation: camshaft chain, drive sprockets and tensioner blades

1 Check the camshaft drive chain for wear and chipped or broken rollers and links. The cam chain operates under almost ideal conditions and unless oil starvation or prolonged tension

adjustment neglect has occurred, it will have a long life. If excessive wear is apparent, or cam chain adjustment has been difficult to maintain correctly, renew the chain.

2   The chain is tensioned via a steel-backed rubber blade by means of a spring loaded plunger. In addition, there is a second blade at the front of the cam chain tunnel which acts as a guide, and a cast bridge piece fitted between the camshaft sprockets which supports the chain upper run. If the rubber material of the guides or blades is worn, the damaged component should be renewed. Extreme wear may indicate a worn drive chain.

2   The cam chain drive sprockets are secured directly to the camshaft ends and in consequence are easily renewable if the teeth become hooked, worn, chipped or broken. The lower sprocket is integral with the crankshaft and if any of these defects are evident, the complete crankshaft assembly must be

renewed. Fortunately, this drastic course of action is rarely necessary since the parts concerned are fully enclosed and well lubricated, working under ideal conditions.

3   If the sprockets are renewed, the chain should be renewed at the same time. It is bad practice to run old and new parts together since the rate of wear will be accelerated.

4   Each cam sprocket is held in place by single bolt, secured by a tab washer backed by a plain washer. Drive is transmitted by a steel pin passing through the sprocket into the camshaft end boss. Note that there are three holes in the cam sprocket which align with similar holes in the boss. The pin should pass into the centre holes. The synthetic rubber faces of the cam sprockets act as a shock absorber for the cam gear. If the rubber has begun to perish or disintegrate, the sprocket concerned should be renewed.

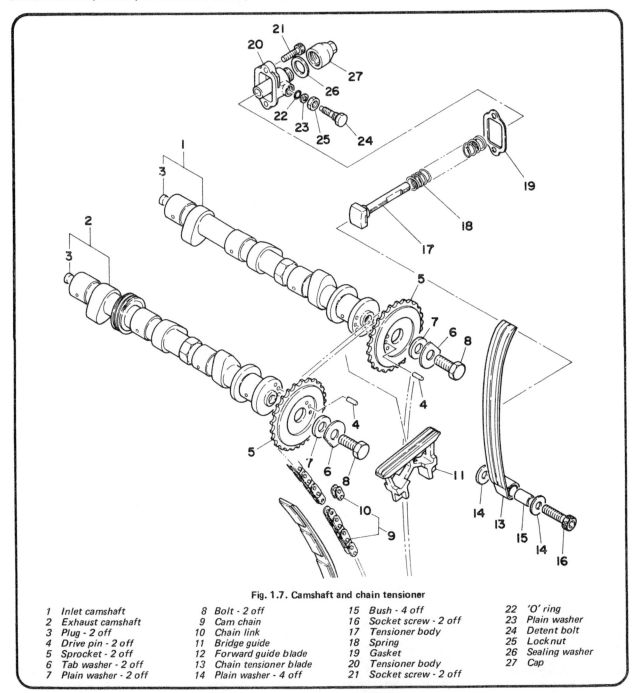

Fig. 1.7. Camshaft and chain tensioner

1	Inlet camshaft	8	Bolt - 2 off	15	Bush - 4 off	22	'O' ring
2	Exhaust camshaft	9	Cam chain	16	Socket screw - 2 off	23	Plain washer
3	Plug - 2 off	10	Chain link	17	Tensioner body	24	Detent bolt
4	Drive pin - 2 off	11	Bridge guide	18	Spring	25	Lock nut
5	Sprocket - 2 off	12	Forward guide blade	19	Gasket	26	Sealing washer
6	Tab washer - 2 off	13	Chain tensioner blade	20	Tensioner body	27	Cap
7	Plain washer - 2 off	14	Plain washer - 4 off	21	Socket screw - 2 off		

28.1 Inspect cam chain drive sprocket for wear

29 Examination and renovation: primary drive chain and sprockets

1   The primary chain is of the Hy-vo or inverted tooth type, and runs around an integral sprocket on the crankshaft and a detachable sprocket secured to the clutch. Check the chain for damaged or loose link plates and pivot pins. This type of chain is very durable and in the normal course of events should have a long service life. Premature wear is unlikely to occur except due to oil starvation. If damage is evident, renew the chain at once. A primary chain which breaks in service will invariably cause extensive engine damage.

2   The service life of the primary chain sprockets is in keeping with that of the chain itself. After considerable use the sprockets may become indented, requiring the renewal of both components. The drive sprocket is an integral part of the crankshaft, and in common with the cam drive sprocket, if wear develops, the crankshaft must be renewed.

30 Examination and renovation: clutch assembly

1   After an extended period of service the clutch linings will wear and promote clutch slip. The limit of wear measured across each inserted plate and the standard measurement is as follows:

Standard	Service limit
Clutch plate thickness 3.0 mm	*2.8 mm*
(0.12 in)	*(0.11 in)*

When the overall width reaches the limit, the inserted plates must be renewed, preferably as a complete set.

2   The plain plates should not show any excess heating (blueing). Check the warpage of each plate using plate glass or surface plate and a feeler gauge. The maximum allowable warpage is 0.05 mm (0.002 in).

3   A damper unit consisting of two rings is fitted behind the final plain plate on the clutch centre boss. The plate is retained by a spring wire clip or a large circlip. If clutch chatter has been experienced the damper ring should be renewed.

4   Check the free length of each clutch spring with a vernier gauge. After considerable use the springs will take a permanent set thereby reducing the pressure applied to the clutch plates. The correct measurements are as follows:

Standard	Service limit
Clutch springs 42.8 mm	*41.5 mm*
(1.685 in)	*(1.634 in)*

5   Examine the clutch assembly for burrs or indentation on the edges of the protruding tongues of the inserted plates and/or slots worn in the edges of the outer drum with which they engage. Similar wear can occur between the inner tongues of the plain clutch plates and the slots in the clutch inner drum. Wear of this nature will cause clutch drag and slow disengagement during gear changes, since the plates will become trapped and will not free fully when the clutch is withdrawn. A small amount of wear can be corrected by dressing with a fine file; more extensive wear will necessitate renewal of the worn parts.

6   The clutch release mechanism attached to the inside of the left-hand engine cover does not normally require attention provided it is greased at regular intervals. It is held to the cover by two cross-head screws and operates on the worm and quick start thread principle.

31 Examination and renovation: clutch shock absorber unit

1   It is unlikely that the shock absorber, which is mounted outboard of the clutch, will require attention during the life of the machine. After considerable mileage has been covered, and during a major overhaul, it may be worthwhile dismantling the unit in order to check the working surfaces of the cam boss, pins and thrust bearing. The shock absorber cam is maintained under tension by four conical plates, retained by a collar and circlip. In order to remove the circlip, the collar must be compressed against the pressure of the springs. Fabricate a U-shaped bridging piece from steel strip, which will clear the splined centre of the shock absorber. Place the unit and the bridge piece in a large vice - see the accompanying photograph - and tighten down sufficiently to release the circlip. Do not compress the plates more than necessary as they may be deformed permanently. Remove the unit from the vice and lift out the four plates and the thrust bearing and side plates.

2   Inspect the three guide pins and the slots in the centre boss which allows the boss to rotate within the fixed limits. The slots should have a smooth curved surface with a uniform radius. The pins should be perfectly round. The efficiency of the units will not be affected greatly until wear of the cam surfaces is pronounced. Slight rippling may therefore be disregarded. Check the four plates for obvious damage. Unfortunately no figures are available by which wear may be evaluated; in practise if the general condition of the shock absorber unit is good, it may be assumed that the plates are fit for further use. Check the thrust bearing; wear or damage to this component will be self-evident.

3   Reassemble the shock absorber unit by reversing the dismantling procedure. When compressing the collar against the plate it is ESSENTIAL that the cam boss **is in the central position relative to the pins.** If this precaution is not taken the distance through which the collar must be compressed to enable installation of the circlip will almost certainly result in damaged plates. Ensure that the circlip is correctly located in the groove before releasing the vice.

32 Examination and renovation: gearbox components

1   The three shafts within the gearbox are supported on journal ball bearings at their right-hand end and caged needle roller bearings on the left. If, on inspection, the bearings or gear pinions are damaged, the bearings must be pulled from place to allow renewal or further dismantling. Examine each of the gear pinions to ensure that there are no chipped or broken teeth and that the dogs on the ends of the pinions are not rounded. Gear pinions with any of these defects must be renewed; there is no satisfactory method of reclaiming them.

2   The gearbox bearings must be free from play and show no signs of roughness when they are rotated. After thorough washing in petrol the bearings should be examined for roughness and play. Also check for pitting on the roller tracks.

3   It should not be necessary to dismantle the gear cluster unless

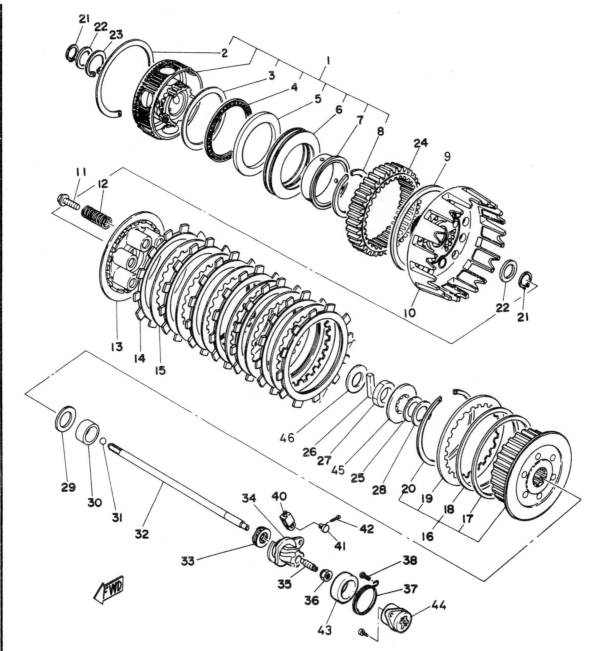

Fig. 1.8. Clutch and shock absorber - component parts

1	Shock absorber (assembly)	12	Spring	25	Conical spring washer	37	Spring
2	Circlip - D, 2D models only	13	Pressure plate	26	Clutch operating crossbar	38	Rivet
		14	Friction plate - 7 off*	27	Nut	39	Screw
3	Thrust washer	15	Plain plate - 6 off	28	Plain washer - not fitted to all models	40	Cable attachment
4	Thrust bearing	16	Centre boss (assembly)			41	Clevis pin
5	Thrust washer	17	Spring seat	29	Washer	42	Split pin
6	Conical spring plate sets	18	Anti-chatter spring	30	Spacer	43	Dust seal
7	Collar	19	Plain plate	31	Steel ball ($\frac{5}{16}$ in)	44	Clutch release worm
8	Circlip	20	Circlip (wire clip on later models)	32	Pushrod	45	Tab washer
9	Shim - some later models only	21	Circlip - 2 off	33	Oil seal	46	Washer - not fitted to some early US models
		22	Thrust washer - 2 off	34	Clutch release mechanism body		
10	Clutch outer drum	23	Circlip	35	Adjusting screw		
11	Spring bolt	24	Primary driven sprocket	36	Nut		

Early models were fitted with 1 moulded cork and 6 moulded resin friction plates, the cork plate being fitted next to the pressure plate. On later models all friction plates were of moulded cork.

31.1a Fashion bridge piece to permit ...

31.1b ... compression of clutch plates in vice

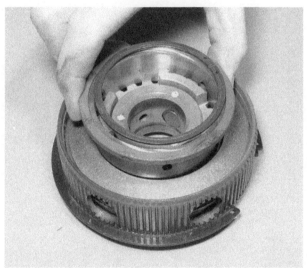

31.1c Remove circlip and collar followed by ...

31.1d ... the two pairs of cone plates

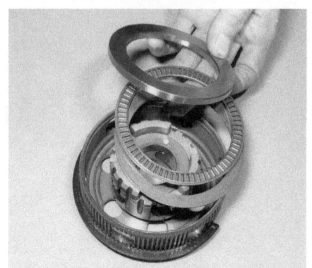

31.3a Check the condition of thrust bearing

31.3b Cam piece must be centralised before refitting circlip

damage has occurred to any of the pinions or a fault has become apparent in a gearbox shaft.

4 The accompanying illustrations show how the clusters are arranged on their shafts. It is imperative that the gear clusters, including the thrust washers and circlips, are assembled in EXACTLY the correct sequence otherwise constant gear selection problems will arise.

5 In order to eliminate the risk of incorrect reassembly make a rough sketch as the pinions are removed. Also strip and rebuild as soon as possible to reduce any confusion which might occur at a later date.

6 Inspect the cam piece in the centre of the middle gear input shaft, and the splined cam piece with which it mates, for wear of the cam profiles. If wear is pronounced, both components should be renewed. The driving cam may be removed from the input shaft after prising out the pressed tin endplate. The cam is retained by a single nut.

7 Check the gear selector rods for straightness by rolling them on a sheet of plate glass. A bent rod will cause difficulty in selecting gears and will make the gearchange particularly heavy.

8 The selector forks should be examined closely, to ensure that they are not bent or badly worn. The case hardened pegs which engage with the cam channels are easily renewable if they are worn. Under normal conditions, the gear selector mechanism is unlikely to wear quickly, unless the gearbox oil level has been allowed to become low.

9 The tracks in the selector drum, with which the selector forks engage, should not show any undue signs of wear unless neglect has led to under lubrication of the gearbox. Check the tension of the gearchange pawl, gearchange arm and drum stopper pawl springs. Weakness in the springs will lead to imprecise gear selection. Check the condition of the gearchange pawl and the pins in the change drum end with which it engages. It is unlikely that wear will take place here except after considerable mileage.

33 Examination and renovation: kickstart shaft assembly and engagement assembly

1 Due to the reliability of the electric start system it is unlikely that the kickstart will be used sufficiently to induce wear. Breakage of the kickstart return spring, which is perhaps the most common fault, can be rectified easily by direct replacement of the damaged spring. The old spring may be pulled off the kickstart shaft, after removal of the two circlips, the backing plate and the spring guide. When fitting a new spring, ensure that the inner turned end is located correctly in the radically drilled hole in the shaft.

2 The kickstart idler double gear which transmits the drive from the kickstart shaft to the engagement mechanism is supported on a caged roller bearing, which, after cleaning in petrol, should be checked for wear and renewed, if necessary.

3 The kickstart engagement mechanism is retained in the upper crankcase half and is supported by a cast outrigger bracket. To remove the mechanism first detach the drive gear from the end of the splined shaft. The gear pinion is retained by a circlip and washer. The support bracket, which is retained by two screws, is a drive fit in the crankcase housing. The centre of the bracket is provided with a threaded hole which may be used to attach a slide hammer to effect removal. If a slide hammer is not available, insert a suitable long bolt into the hole and screw it inwards slowly, so pushing the bracket from position. The blade of a large screwdriver or a similar piece of steel plate should be placed between the screw end and the engagement pinion to prevent damage to the pinion teeth. Any damaged components can now be removed. When refitting the shaft into the casing, note that the tongue of the kickstart clip, attached to the engagement pinion, must engage with the guideway in the casing.

32.6a Middle gear input shaft needle roller bearing and oil seal

32.6b Check condition of cam shock absorber faces

33.3a Kickstart drive gear is retained by a circlip

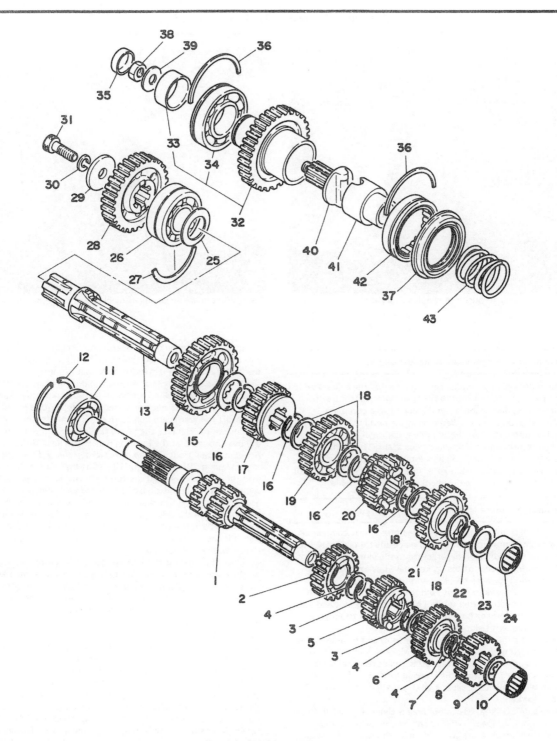

Fig. 1.9. Gearbox components

1	Mainshaft	11	Journal ball bearing	22	Circlip	32 Middle gear input shaft and
2	Mainshaft 4th gear pinion	12	Circlip	23	Washer	gear pinion
3	Circlip - 2 off	13	Layshaft	24	Caged needle roller bearing	33 Collar
4	Washer - 3 off	14	Layshaft 1st gear pinion	25	Washer	34 Journal ball bearing
5	Mainshaft 3rd gear pinion	15	Washer	26	Journal ball bearing	35 Plug
6	Mainshaft 5th gear pinion	16	Circlip - 4 off	27	Circlip	36 Half clip - 2 off
7	Circlip	17	Layshaft 4th gear pinion	28	Final output gear	37 Double lip oil seal
8	Mainshaft 2nd gear pinion	18	Shim - 4 off	29	Washer	38 Nut
9	Washer	19	Layshaft 3rd gear pinion	30	Spring washer	39 Plain washer
10	Caged needle roller	20	Layshaft 5th gear pinion	31	Socket bolt	40 Drive cam piece
	bearing	21	Layshaft 2nd gear pinion			41 Drive cam piece
						42 Caged roller bearing
						43 Shock absorber spring

33.3b Use a screw to push out support bracket

33.3c Manoeuvre engagement mechanism from the casing

34 Examination and renovation: middle gear box

1   The middle gearbox which contains the 90° bevel gears which transmit the gearbox output to the final drive shaft is a self-contained unit, detachable from the engine as a complete assembly. If excessive wear develops in the bevel box or if any internal components fail, the complete, undismantled unit should be returned to a Yamaha service agent. The manufacturer advises strongly that the bevel box should be inspected and overhauled by qualified and experienced operators only, and for this reason no service or adjustment figures are available.

35 Engine reassembly: general

1   Before reassembly of the engine/gearbox unit is commenced, the various component parts should be cleaned thoroughly and placed on a sheet of clean paper, close to the working area.
2   Make sure all traces of old gaskets have been removed and that the mating surfaces are clean and undamaged.
One of the best ways to remove old gasket cement is to apply a rag soaked in methylated spirit. This acts as a solvent and will ensure that the cement is removed without resort to scraping and the consequent risk of damage.
3   Gather together all of the necessary tools and have available an oil can filled with clean engine oil. Make sure all new gaskets and oil seals are to hand, also all replacement parts required. Nothing is more frustrating than having to stop in the middle of a reassembly sequence because a vital gasket or replacement has been overlooked.
4   Make sure that the reassembly area is clean and that there is adequate working space. Refer to the torque and clearance settings wherever they are given. Many of the smaller bolts are easily sheared if over-tightened. Always use the correct size screwdriver bit for the crosshead screws and never an ordinary screwdriver or punch. If the existing screws show evidence of maltreatment in the past, it is advisable to renew them as a complete set.

36 Engine reassembly: replacing the gearbox and the gear selector components

1   Insert the gearchange drum through the gearbox left-hand wall so that the end of the drum projects into the gearbox chamber. Replace the drum stopper quadrant drive pin and then fit the stopper quadrant so that it engages with the pin. Refit the retaining circlip.

2   Install the change drum locating bolt together with the tab washer, the edge of which should be bent up against one of the bolt head flats. Fit the drum stopper plunger, detent spring and housing bolt. Replace the neutral indicator switch after checking that the sealing washer is in good condition. Take care when tightening the neutral switch, as it may quite easily shear off.
3   Invert the crankcase half so that the upper mating surface is downwards. Replace the layshaft centre bearing together with the bearing cap and the locating half-clip. The clip should be fitted into the groove in the cap. Tighten the cap nuts evenly to a torque of 1.8 - 2.2 kg m (13 - 16 ft lbs). Insert the layshaft into the gearbox through the tunnel in the casing wall. At this stage the layshaft must have been fitted with the 4th gear, 3rd gear, 2nd gear and 5th gear pinions and the interposed circlips and shims. See the appropriate line drawing for the assembly sequence. With the end of the layshaft projecting into the gearbox. Fit the 1st gear pinion, which should be flanked by two spacer washers of equal size.
4   Place the final output gear between the centre bearing and gearbox wall and move the layshaft over until the shaft engages with the pinion's splines. Secure the completed layshaft by refitting the end bolt, large washer and spring washer to the primary drive end of the shaft. Use a similar technique to that employed during dismantling for locking the shaft whilst the bolt is tightened.
5   Place the crankcase lower half on its right-hand side so that layshaft tunnel is upright. Fit the endfloat shim onto the layshaft so that it rests against the 2nd gear pinion retaining circlip. Check that the large O-ring is in place on the layshaft bearing housing and position the housing, complete with bearing, over the tunnel. The gearchange shaft support arm on the housing must line up with the shaft hole in the casing when the housing is finally in position. If required, the gearchange shaft may be inserted temporarily to aid this operation. Using a rawhide mallet tap the housing fully home, ensuring that it stays square as it is driven into place. Fit and tighten the four housing screws and withdraw the gearchange shaft.
6   Position the crankcase lower half, upper face downwards. Install the two lower selector forks so that their fingers engage with the channels in the 4th and 5th gear pinions, and so that their guide pins engage with the channels in the change drum. During this operation the change drum should be placed in the neutral position. This may be found by rotating the drum as far as possible in an anticlockwise direction - viewed from the pin end - and then turning it back one notch. Insert the selector fork rod from the left so that it passes through the gearbox wall and engages with the two selector forks. Place the plain washer on the rod before pushing it fully home, and then fit the securing

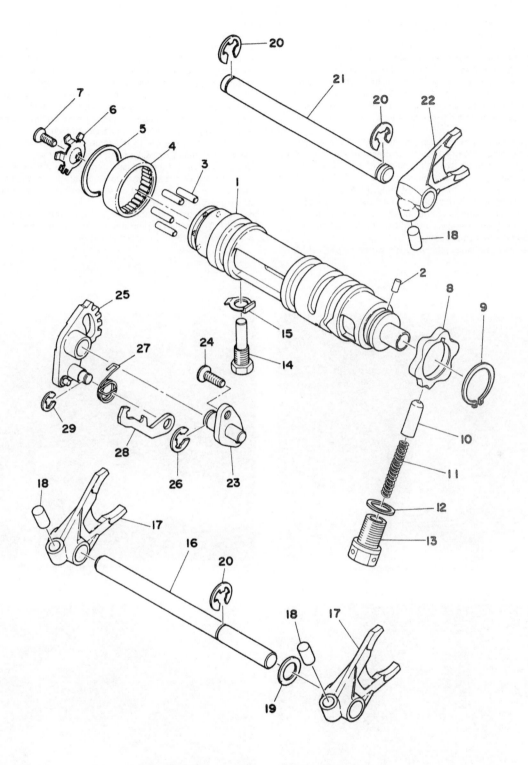

Fig. 1.10. Gear selector mechanism - component parts

1	Gearchange drum	8	Drum stopper quadrant	15	Lockwasher
2	Drive pin	9	Circlip	16	Lower selector fork rod
3	Pin - 4 off	10	Detent plunger	17	Selector fork - 2 off
4	Caged roller bearing	11	Spring	18	Guide pin - 3 off
5	Circlip	12	Sealing washer	19	Washer
6	Pin retainer plate	13	Detent bolt/housing	20	'E' clip
7	Countersunk screw	14	Location bolt	21	Upper selector fork rod

22	Selector quadrant
23	Quadrant pivot shaft
24	Countersunk screw - 2 off
25	Selector quadrant
26	'E' clip
27	Pawl spring
28	Change pawl
29	'E' clip

E clip so that the washer lies between the E clip and the support lug on the centre bearing cap.

7   Invert the crankcase lower half so that it is the right way up. Fit the completed mainshaft as a single item, with a bearing on each end. Ensure that the journal ball bearing retaining circlip is located correctly in the groove in the casing. Position the upper selector fork so that it engages with the 3rd gear pinion and the change drum. Insert the selector fork rod and fit the retaining circlips on both ends. Position a new clutch pushrod oil seal in the casing.

8   Install the middle gear input shaft complete with the shock absorber campiece, both bearings and the oil seal. The journal ball bearing is located by two half clips. Each half clip should fit into both casings, ie. an imaginary line drawn across both ends of each clip should be vertical. The oil seal, which is of a special type having two garter springs, should be placed on the shaft with the locating shoulder towards the bearing, so that the shoulder fits into the casing groove. Lubricate the sealing lip before fitting the seal to the shaft.

9   Replace the gearchange selector quadrant on its pivot shaft, securing the component with the washer and circlip. The pawl arm must engage with the pins in the end of the change drum. Lubricate the end of the gearchange shaft so that when it passes through the shaft oil seal, the sealing lip will not become damaged. Insert the shaft from the primary drive side, pushing it through until the endfloat washer may be fitted on the left-hand side. Push the shaft fully home, ensuring that the punch mark on the gearchange shaft arm aligns with the punch mark on the selector quadrant. If alignment is not ensured **correct gear selection will not be possible.** Fit the shaft securing circlip.

10 Before continuing with reassembly, the gear selection mechanism must be checked and if necessary adjusted. Place the change drum in the second gear position. In this position the scribed mark on the selector quadrant pawl arm must align with a similar mark scribed on the change drum end plate. If adjustment is required, loosen the locknut on the main gearchange shaft arm and rotate the eccentric bolt either to the left or right. This will bring the two marks into alignment. After adjustment, tighten the locknut and recheck.

36.1a Replace the stopper quadrant drive pin and ...

36.1b Insert change drum and fit stopper and circlip

36.2a Bend up tab washer to secure change drum locating bolt

36.2b Replace drum detent mechanism and ...

36.2c ... the neutral indicator switch

36.3a Fit the layshaft centre bearing and clip to the cap

36.3b Replace the assembly in the casing

36.3c Insert layshaft and fit 1st gear pinion and washer

36.4a Replace the layshaft end bolt and washers ...

36.4b ... using a sprag to hold the shaft whilst tightening

36.5a Install the end shim on layshaft

36.5b Do not omit 'O' ring on bearing housing

36.6a Selector forks are not interchangeable

36.6b Do not omit washer on shaft

36.7a Position the completed mainshaft in the casing

36.7b Install the selector rod and fork and ...

36.7c ... secure the rod by means of the 'E' clips

36.7d Position clutch push rod seal

36.8a Install bearing retaining half clips and ...

36.8b ... replace middle gear input shaft as a unit

36.9a Fit gear change selector quadrant and ...

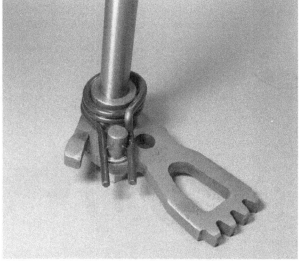

36.9b ... position the centraliser spring as shown

36.9c Gearchange shaft dot and quadrant dot must align

36.9d Fit washer and 'E' clip to secure shaft

36.10a Scribed marks MUST align as shown

36.10b Use adjuster screw to align marks

37 Engine reassembly: replacing the crankshaft and joining the crankcase halves

1   Install the main bearing shells into the upper and lower crankcase halves, ensuring that each is positioned correctly with both ends flush with the mating surface. The bearing shell which controls the crankshaft endfloat must be fitted in the upper casing half, in the second housing from the left (viewed from above).

2   Lubricate the lower shells thoroughly, using clean engine oil. The crankshaft, complete with connecting rods, may now be lowered into position.

3   Check that all the shafts within the lower crankcase half are correctly positioned. Fit the small O ring into the mating surface of the lower casing between the camshaft drive end of the crankshaft and the gearbox mainshaft. Lubricate the main bearing journals and the areas around the big-ends. Insert the two crankcase locating dowels. Apply good quality non-hardening gasket compound to the mating surface of the upper crankcase half. The manufacturers recommend Yamabond No. 4 sealant.

4   The upper crankcase half can now be lowered into position. It is helpful at this stage to have an assistant at hand to feed the three connecting rods through the crankcase mouths as the upper casing is lowered. Fit the upper crankcase bolts, screwing

them in finger tight only. Note that bolts numbered 5, 6, 7 and 8 are fitted with copper washers. Invert the crankcase and fit the lower bolts.

5   The crankcase bolts should be tightened following the numerical sequence, commencing at bolt No. 1 and continuing through to the unnumbered bolt which should be considered as No. 24. **Do not omit the hidden bolt** located in the primary drive compartment. The bolts should be tightened in two stages to the following torque settings:

	1st stage	2nd stage
8 mm bolts	1.0 kg m (7 ft lb)	2.0 kg m (14 ft lb)
10 mm bolts	2.0 kg m (14 ft lb)	3.7 kg m (27 ft lb)

6   After tightening the crankcase bolts, check that the crankshaft and gearbox shafts still rotate freely. Temporarily refit the gear selector lever and attempt to select each gear in turn. If any difficulty is encountered in selection, now is the time to find out rather than at a later stage when more dismantling work would be needed to rectify the fault. Problems with selection may be due to incorrectly assembled gear clusters or a mistimed selector quadrant. This and the selector alignment should be checked and if necessary adjusted.

37.1 Install main bearing shells; projections must locate with casing recesses

37.2a Lower the crankshaft into position

37.2b DO NOT OMIT this 'O' ring

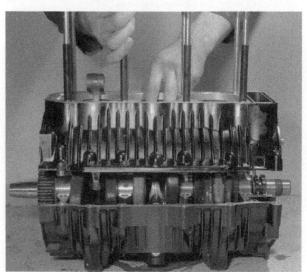

37.4a Feed connecting rods through crankcase mouths

37.4b Refit clutch mechanism boot, bracket and ...

37.4c ... bridge piece and cable clip

37.5 Don't forget the hidden crankcases bolt

38 Engine reassembly: replacing the kickstart assemblies

1   If the kickstart engagement mechanism was removed it should be refitted now into the upper casing recess. Position the shaft, engagment pinion and kickstart clip in the casing. The clip tongue must engage with the groove in the case. Place the support bracket over the shaft and tap it home squarely so that the two screw holes align. Fit and tighten the three retaining screws. Replace the driven gear on the end of the shaft, securing it with the washer and circlip.

2   Place the kickstart idler gear thrust washer and caged roller bearing on the stub end of the middle gear input shaft. Lubricate the bearing. The kickstart idler gear must be fitted simultaneously, with the completed kickstart shaft. When both are in position, grasp the outer turned end of the return spring with a pair of pliers and tension it in a clockwise direction, until the hook may be anchored on the cast anchor lug in the casing. Temporarily refit the kickstart lever and check that the system functions correctly. Fit the idler gear retaining circlip.

38.1a Insert kickstart mechanism into casing and ...

38.1b ... fit the support bracket and screws

38.1c Replace the drive gear, secured by the washer and circlip

38.2a Install kickstart shaft and idler gear simultaneously

38.2b Tension return spring, to anchor on lug

38.2c Fit the idler gear retaining circlip

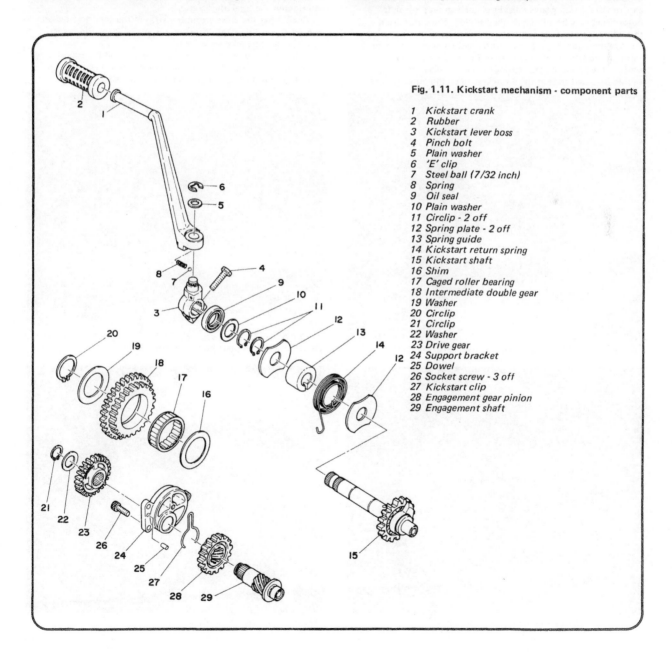

Fig. 1.11. Kickstart mechanism - component parts

1 Kickstart crank
2 Rubber
3 Kickstart lever boss
4 Pinch bolt
5 Plain washer
6 'E' clip
7 Steel ball (7/32 inch)
8 Spring
9 Oil seal
10 Plain washer
11 Circlip - 2 off
12 Spring plate - 2 off
13 Spring guide
14 Kickstart return spring
15 Kickstart shaft
16 Shim
17 Caged roller bearing
18 Intermediate double gear
19 Washer
20 Circlip
21 Circlip
22 Washer
23 Drive gear
24 Support bracket
25 Dowel
26 Socket screw - 3 off
27 Kickstart clip
28 Engagement gear pinion
29 Engagement shaft

39 Engine reassembly: replacing the clutch assembly

1   Refit to the mainshaft, the spacer and washer, followed by the clutch centre boss, the plain washer (where fitted), the conical spring washer (with its convex surface outwards), the tab washer and the nut. Tighten the nut using the long 32 mm box spanner and applying the same technique for preventing rotation of the shaft as was used during dismantling. Bend up the side of the tab washer to secure the nut.
2   Fit the clutch plates one at a time, commencing with a friction plate followed by a plain plate and then replacing the remainder alternately. Each plain plate has a small tab projecting from the outer edge. Fit the first plate so that the tab aligns with one of the dot marks on the outer face of the centre boss. The second should be fitted with the tab aligned with the next dot around the face. Continue fitting the plates so that the tabs are spaced out evenly, aligned with the dots, in a clockwise direction.
3   Before fitting the clutch pressure plate and springs, the ears of the friction plates must be aligned so that the clutch outer drum may be subsequently installed. The outer drum should be used to align the plates. After alignment is complete, remove the outer drum and insert the clutch operating crossbar through the slot in the clutch shaft. Fit the heavy plate washer, the pressure plate, springs and bolts. The pressure plate should be fitted so that the arrow marks on the outer surface

align with the similar marks on the clutch centre boss face. Tighten the spring bolts evenly, until they are tight.
4   Fit the circlip onto the shaft, followed by the plain washer and finally the clutch outer drum. Secure the drum by means of the washer and small circlip which are positioned side by side on the end of the mainshaft.

40 Engine reassembly: replacing the primary drive assembly and clutch shock absorber

1   Place the primary driven sprocket over the clutch shaft so that the more pronounced raised shoulder faces outwards. Position the primary drive chain so that it meshes with both sprockets. Insert the clutch shock absorber into the driven sprocket and push it fully home, then refit the large circlip to retain the shock absorber.
2   Replace the outrigger support complete with bearing so that it locates with the four hollow dowels in the casing. Fit and tighten the retaining screws.
3   Check that the crescent rubber strip in the support bracket is in good condition and is securely in place. Place the thrust washer over the kickstart shaft.
4   Place a new gasket on the mating surface and fit the primary drive casing. Replace the retaining screws, but do not tighten them fully until the alternator has been refitted.

39.1a Position the spacer collar and washer on clutch shaft

39.1b Inner plain plate held by spring ring

39.1c Fit clutch centre nut and washer and ...

39.1d ... lock clutch to tighten the nut

39.2a Replace the crossbar and heavy washer

39.2b Install the clutch plates alternately

39.3a Replace the clutch pressure plate so that ...

39.3b ... the arrows on the plate align with those on the centre

39.3c Fit and tighten the springs and bolts

39.4a Install the small circlip and washer and ...

39.4b ... refit the clutch outer housing and ...

39.4c ... secure the housing by means of the washer and circlip

40.1a Position Hy-Vo chain and driven sprocket

40.1b Insert the completed shock absorber unit and ...

40.1c ... retain by means of the final circlip

40.2 Fit the outrigger bearing casting

40.3 Check that the rubber strip is in position

40.4 Use a new gasket on the primary drive casing

41 Engine reassembly: replacing the oil pump and sump

1   Check that the rubber sealing ring is in place in the base of the oil pump main body. Place the pump in position so that it locates with the two hollow dowels. Fit and tighten the oil pump screws evenly. Refit the gauze filter funnel and the filter screen and gasket.
2   It is **ESSENTIAL** that the oil pump is not fitted dry as a problem may be encountered in self-priming when the engine is first started after an overhaul. To overcome this problem, pour clean engine oil into the pick-up orifice and rotate the oil pump pinion in an anti-clockwise direction until oil is seen emerging from the outlet.
3   Lubricate all the accessible gearbox shafts and bearings with a copious quantity of clean engine oil. Place a new gasket on the sump mating surface and refit the sump. Insert and tighten the retaining screws evenly.

41.1a Prime the oil pump before fitting into casing

41.1b Replace the pick-up funnel and ...

41.1c ... the gauze filter screen and gasket

42 Engine reassembly: replacing the middle gearbox

1   Place the engine in an upright position, resting on the sump. Position a new gasket on the middle gear face of the crankcase and check that the single locating dowel is fitted to the middle gearbox mating surface. Fit the shock absorber spring and campiece to the splined middle gear shaft.
2   Position the middle gear case against the crankcase and insert the retaining screws. Tighten the screws evenly, a little at a time, whilst the cam spring is being compressed. Check that the middle gear campiece has engaged correctly with the campiece in the input shaft by rotating the final drive output flange. The gear case retaining bolts should be tightened to a torque of 2.0 - 2.5 kg m (14 - 18 ft lbs).

43 Engine reassembly: replacing the cam chain, guides and starter motor/oil pump drive gears

1   Replace the oil feed cross-pipe in the left-hand gearcase. Check that the sealing washers on each banjo union are in good condition. Make sure when fitting the pipe that the banjo unions do not foul the crankcase; during 1978 the pipe union was modified to prevent this. Do not overtighten the banjo bolts as they are not only hollow but cross-drilled and are therefore easy to shear.
2   Replace the two cam chain guide blades, noting that each is fitted with a pivot bush and a plain washer either side of the lower end eye. Tighten the bolts fully and check that both blades are free to move. The blade with the three protruding blocks on the reverse face should be fitted at the front of the engine.
3   Insert one end of the cam chain through the top of the crankcase and run the chain round the drive sprocket and up out of the casing. The chain ends may be draped over the guide blades at this stage.
4   Lubricate the bearing boss on the starter motor/oil pump drive shaft assembly and insert the complete unit into the casing so that the small gear pinion on the shaft end engages with the oil pump driven pinion. Do not omit from the shaft the shim between the flanged bearing face and the starter clutch unit, or from the outside of the clutch unit.
5   Install the starter motor intermediate gear pinion together with the hollow shaft and locating circlip. The greater length of shaft should point away from the gear pinion.

6   Fit the spur gear to the end of the crankshaft so that the pins engage with the drive splines. Replace the plain washer and gear pinion retaining nut. Lock the crankshaft in the normal manner by placing a close fitting bar through the small end eye of one connecting rod. The spur gear nut should be tightened to a torque setting of 8.0 - 12 kg m (58 - 87 ft lbs), ie. very tight!
7   The clutch push rod and the small steel ball which precedes it should be fitted into the hollow clutch shaft (gearbox mainshaft) before replacing the left-hand engine cover. Grease the pushrod before insertion and note that it should be fitted with the plain shouldered end towards the left.
8   Grease the clutch operating mechanism in the left-hand engine cover, and lightly lubricate the contact breaker oil seal. Fit a new gasket to the cover face, ensuring that the two locating dowels are in place. Replace the cover and tighten the screws, noting that the centre and right-hand screws in the lower run also retain two wiring clips.

42.1a Position the spring and cam piece on the input shaft

42.1b A new gasket should be used

43.1 Replace the oil cross feed pipe

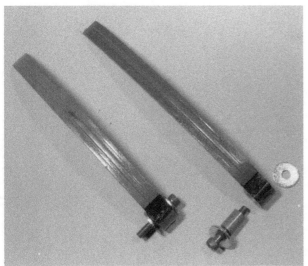

43.2a Cam chain guides - general view

43.2b Check guides can move after tightening bolts

43.3 Feed cam chain around the drive sprocket

43.4a Lubricate the starter clutch bush and ...

43.4b ... insert clutch unit with washer on shaft

43.4c Fit shim to outside of clutch unit

43.5 Install starter intermediate gear pinion and shaft

43.6a Oil pump drive gear pins must engage with drive splines

43.6b Fit the plain washer and centre nut

43.7a Insert the steel ball into the hollow mainshaft ...

43.7b ... and fit the clutch pushrod

43.8 Grease the clutch lifting mechanism

44 Engine reassembly: replacing the ATU and contact breaker assembly

1 Insert the ATU drive pin in the end of the crankshaft and position the timing unit, so that the drilling in the rear face engages with the pin. Place the complete contact breaker assembly mounting plate in the casing so that the wiring grommet locates with the relieved portion of the casing wall. If alignment marks were made on the stator plate and an adjacent part of the casing before dismantling was commenced, realign the marks and fit and tighten the three retaining screws. The static ignition timing is now correct. After assembly is complete, the timing should be checked using a stroboscope. Where alignment marks were not made, static timing and then stroboscopic timing should be carried out as described in Chapter 3, Section 7.
2 Replace the engine turning hexagon and the centre bolt which which secures both it and the ATU.
3 At this juncture in reassembly do not replace the contact breaker cover as access to the engine rotation hexagon is still required.

44.1b ... fit the ATU so that the pin locates correctly

44.1a Replace the ATU drive pin and ...

45 Engine reassembly: replacing the pistons and cylinder block

1 Before continuing with assembly it is worthwhile making up a simple tool to hold the cam chain in a vertical position and so free the hands when replacing the cylinder barrel and cylinder head. Using a length of welding rod or a straightened coathanger, form a tool similar to the item shown in the accompanying photograph. The tool may be installed so that the closed end rests on the crankshaft adjacent to the cam chain drive sprocket and one end of the chain hooks, under tension, over each of the upper arms.
2 Pad the crankcase mouths with clean rag to prevent the ingress of foreign matter during piston and cylinder barrel replacement. It is only too easy to drop a circlip while it is being inserted into the piston boss, which will necessitate a further strip down for its retrieval.
3 Replace the rings on each piston, noting that the two compr-

ession rings have different profiles. The outer diameter of the upper ring is slightly tapered from the top, with gently chamfered edges. The second compression ring has a heavily bevelled top edge. Both rings should be fitted with the stamp mark upwards. When fitting the lower ring (oil control) ensure that the gaps in the outer rails are staggered from one another. another.
4 Fit the pistons onto their original connecting rods, with the arrow embossed on each piston crown facing forwards. If the gudgeon pins are a tight fit in the piston bosses, warm each piston first to expand the metal. Do not forget to lubricate the gudgeon pin, small end eye and the piston bosses before reassembly.
5 Use new circlips. NEVER re-use old circlips. Check that each circlip has located correctly in its groove. A displaced circlip will cause severe engine damage.
6 Fit a new cylinder base gasket and a new cam chain tunnel sealing ring. The flat side of the sealing ring must face downwards. Install the two oil passage O rings and hollow dowels to the right-hand end of the crankcase cylinder mating surface. The flat faces of these seals must face upwards. Replace the two smaller hollow locating dowels.
7 Rotate the engine so that No. 1 piston is at TDC. Position the piston rings so that their end gaps are out of line with each other. Refitting of the cylinder block can be facilitated by the use of a piston ring clamp placed on each piston. This is by no means essential because the cylinder bore spigots have a good lead-in and the rings may therefore be hand-fed into the bores. Whichever method is adopted, an assistant should be available to guide the pistons.
8 Position a new sealing ring around each cylinder bore spigot and lubricate the bores thoroughly with clean engine oil. Carefully slide the cylinder block down the holding down studs until the pistons enter the cylinder bores; keeping the pistons square to the bores ease the block down until the piston clamps are displaced. Remove the piston ring clamps and the rag padding from the crankcase mouths and push the cylinder block down onto the base gasket.
9 When lowering the cylinder block, the cam chain and guide blades must be fed through the chain tunnel. If the previously mentioned special holding tool has been fitted, the guide blades may be held by an elastic band or a wire clip. Where the tool is not employed, the chain should be hooked up, using two lengths of wire. If the latter method is used, ensure that both chain ends are secured once the cylinder block is in place.

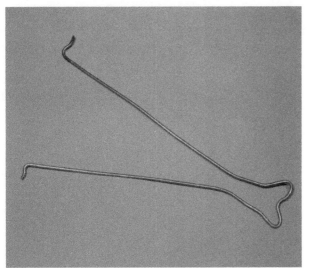

45.1a Fabricate simple tool from wire to ...

45.1b ... support cam chain during further assembly

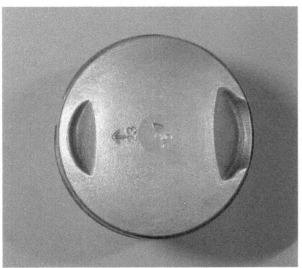

45.4a Arrow on piston crown must face forwards

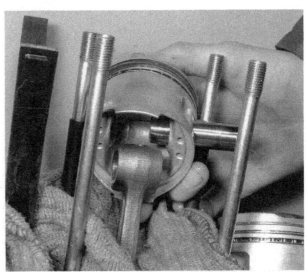

45.4b Lubricate gudgeon pins before assembly

45.5 Always use NEW circlips

45.6a Fit a new cam chain tunnel seal and ...

45.6b ... new base gasket and dowel seals

45.6c Dowel seal must be fitted with curved surface downwards

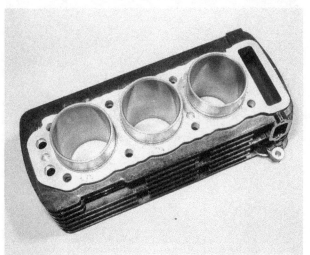

45.8 Place new 'O' rings on each cylinder sleeve spigot

45.9 Feed cylinder block onto pistons

46 Engine reassembly: replacing the cylinder head

1   Before the cylinder head can be refitted the valves, springs
and cam followers must be replaced by reversing the dismantling
procedure. Place a new valve guide seal onto each guide top.
Lubricate the valve stems thoroughly. When fitting the springs do
not omit the lower sealing washer. Note that each spring has
variable pitch coils. The springs MUST be fitted so that the
more widely spaced coils are at the top. After fitting the valve
collets and releasing the spring compressor strike smartly the top
of each valve stem with a hammer. This will ensure that the
collets are seated correctly. Lubricate the cam followers and
install them, ensure that each is returned to its original location.
Fit also the adjuster pads in their original locations.
2   Fit a new cylinder head gasket to the top of the cylinder
block, with the gasket tab facing forwards. Install the two oil
passage seals (flat side uppermost) and dowels, the two smaller
locating dowels and the cam chain tunnel sealing ring. At this
stage the chain holder should be removed and the chain fed

through the tunnel by means of separate wires.
3   Lift the cylinder head and place it carefully in position on the
cylinder block. Lubricate the threads of the cylinder head bolts
before fitting and note that the two central bolts on the exhaust
side of the head are fitted with heavier plain washers than the
remainder.
4   As with the crankcase bolts, the cylinder head nuts and bolts
must be tightened in the correct sequence in two stages.
Cylinder head bolts torque settings:

	1st stage	2nd stage
8 mm bolt (Nos. 9, 10)	1 kg m (7 ft lbs)	2.0 kg m (14 ft lbs)
10 mm mut (Nos. 1 - 8)	1.5 kg m (11 ft lbs)	3.5 kg m (25 ft lbs)

Refer to the accompanying diagram for the correct sequence.
5   Replace the cam gear oil feed pipe at the rear of the cylinder
block. Check that the two sealing rings on each banjo union
are in good condition and do not overtighten the banjo bolts.

46.1a Fit a new valve guide seal and ...

46.1b ... insert the valves after lubricating the stems

46.1c Fit the spring lower seat

46.1d Widely spaced spring coil MUST be at top

46.1e Replace the spring upper seat (collar)

46.1f Compress the spring to fit the collets

46.1g Correct location of the collets is essential

46.2a Lubricate and refit the cam followers and adjuster pads

46.2b Always use new gasket and seals at cylinder head

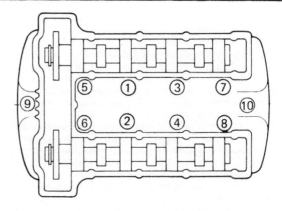

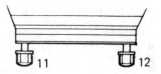

Fig. 1.12. Cylinder head tightening sequence

47 Engine reassembly: replacing the cams, and timing the valves

1   The camshafts must be fitted complete with the camshaft sprockets. Place No. 1 piston at TDC, if this has not already been done. If required, move the cam chain around the lower sprocket so that equal lengths of chain extend from the cylinder head on both sides. Resecure the chain to prevent it falling down into the gear case.

2   Lubricate the exhaust cam bearing surfaces and the cam bearing journals. Position the cam on the cylinder head so that the valve timing index dot on the camshaft boss is at 12 o'clock. Now turn the cam clockwise 1/6 of a turn, to bring the camshaft journal more closely towards the bearing surfaces. Fit the four numbered cam bearing caps with the arrow on each pointing to the left. The No. 1 cap is fitted next to the sprocket. Replace the retaining nuts and washers and tighten the nuts evenly from left to right.

3   Fit the inlet camshaft using a similar procedure. The shaft should be turned anti-clockwise 1/6 of a turn after being positioned at the 12 o'clock mark.

4   Using the hexagon on the centre of each camshaft turn the exhaust camshaft anti-clockwise 1/6 of a turn, and the inlet camshaft clockwise 1/6 of a turn so that the timing dots align with the index marks on the cam bearing caps nearest the sprockets. Check that No. 1 cylinder is at TDC by viewing the ATU base plate through the inspection aperture in the contact breaker baseplate. The T mark to the right of the 1F mark must be **exactly** aligned with the index mark on the casing. Do not turn either camshaft other than the amount stated or an opening valve may strike the piston top.

5   Pad the area immediately adjacent to the camshaft sprockets with clean rag to prevent the cam chain link or side plate from falling into the chain tunnel during fitting. Mesh the two ends of the cam chain over the sprockets and insert the new chain link from the rear. The old link must **NOT** be reused. Fit the link side plate. The ends of the link pins must be rivetted over to retain the link side plate securely. This operation must be done correctly or the link may separate with the engine running causing extensive damage. Use a centre punch with the end ground at such an angle that it will spread the ends of the link pins outwards from the centre. The rear side of the link **must be supported** against the side thrust during rivetting. A 2 lb

hammer with a flat face or a similar block of steel is ideal. If there is any doubt concerning the rivetting procedure, seek the advice of a qualified mechanic.

6   Replace the upper chain guide and fit the chain tensioner, together with the gasket. The tensioner relies on the plunger spring being in good condition to accurately adjust the pressure on the guide blade, whenever the adjustment is reset. If the tensioner has been in service for an extended length of time, renew the spring as a precaution.

7   Adjust the cam chain as follows. Remove the tensioner cap and slacken the plunger detent bolt and locknut. Rotate the engine forwards by applying a spanner to the turning hexagon. As the engine is turned, the tensioner plunger will move in and out as the cam chain tension alters. Find the position at which the plunger is at its innermost point. Tighten the detent bolt and locknut and fit the tensioner cap.

8   Before continuing, rotate the engine until No. 1 piston is at TDC and recheck the valve timing.

---

48 Engine reassembly: checking and adjusting the valves

---

1   The clearance between each cam and cam follower is adjusted by means of removable pads which are seated in the top of each cam follower. As can be seen by the relevant tables in the routine maintenance section at the beginning of the manual, 25 pads are available, each of which is of slightly different thickness. Rather than acquiring the complete set of pads, which in any case may be duplicated in part by those already in the engine, it is suggested that the required thickness of the pads be ascertained and the correct pads purchased as needed. It is also likely that two or more pads of the same size will be required. Where the cylinder head components have not been disturbed ie. the valves have not been removed for renewal or attention, valve clearance adjustment should be made now and then at the next applicable service interval. Where the valve components have been renewed or refurbished, adjust the clearances at this stage and then again after the engine has covered approximately 500 miles and has bedded down.

2   To check the valve clearance rotate the engine until the cam lobe is pointing away from the cam follower. Insert a feeler
a   gauge and check that the clearance is as follows:

Inlet valves   0.16 - 0.20 mm (0.006 - 0.008 in)
Exhaust valves   0.21 - 0.25 mm (0.008 - 0.010 in)

If the clearance is incorrect, the adjuster pad must be removed to ascertain which pad to install to obtain the correct clearance. A special service tool is available which may be secured to the

cylinder head adjacent to the valve in question, and so hold down the cam follower whilst the pad is being removed. The tool can be fitted only when the valve has already been depressed. Rotate the engine until the valve is fully open. Fit the holding tool so that the tongue is contacting the raised edge of the cam follower but not touching the pad. The securing screw must be tightened fully to prevent the tool rotating. Before fitting the tool, rotate the cam follower so that the slot in the edge is opposite the tool; this will aid pad removal. The engine must now be turned so that the cam lobe is in the clearance checking position permitting the pad to be dislodged. It is **absolutely vital** that the cam lobe is **not allowed to touch the holding tool** as the resultant force can break the cylinder head casting. For this reason the exhaust cam must be turned **anti-clockwise only,** and the inlet cam **clockwise only** (viewed from the left). Hook out the adjustment pad and note the number stamped on the underside. A magnetic rod or magnetised screwdriver aids this operation considerably.

3   The correct pad needed to restore the clearance to within the specified range may be found by referring to the relevant table, using the installed pad number and the clearance measured. Select the correct adjuster pad and install it in the cam follower. Rotate the engine until the cam lobe is in contact with the pad and then remove the holding tool. The pad should be positioned with the identification number downwards. Rotate the engine a number of times to ensure that the pad has seated and then recheck the clearance. If the clearance is not satisfactory the adjustment procedure should be repeated, using the newly installed pad number as a guide.

4   If the valve holding tool is not available, an alternative procedure may be adopted, for which an easily constructed cam follower holding tool must be fashioned. The tool may be made from a length of steel plate approximately 6 x ¾ x ¼ in relieved at one end to fit between the edge of the cam follower and the camshaft. In addition, the tool end must be bent so that it will clear adjacent cylinder head castings.
Rotate the engine until the cam lobe of the valve to be attended to is in the clearance checking position. Using a stout, long-handled screwdriver between the cam heel and the adjuster pad, lever the valve into the open position and insert the tool. Ensure that the tool is clear of all castings and securely placed on the edge of the cam follower before releasing the leverage from the screwdriver. **Extreme caution should be used** when applying this method because slippage of either lever may result in damage to the components. Use the correct tool if at all possible.

5   After checking the clearances, refit the camshaft cover, using a new gasket. Insert and screw in finger tight all but the air scoop brackets retaining screws. Do not tighten the cover screws until the brackets have been fitted.

47.2a Camshaft caps must be fitted with arrow to the left

47.2b Fit camshafts with marks as shown

47.4 Turn both camshafts so dots align with projections

47.5a Connect the chain ends using a NEW link

47.5b Fit the link plate and rivet the pin ends

47.6 Replace the cam chain tensioner and adjust

48.5 Install the cam box end seals before fitting cover

49 Replacing the engine/gearbox unit in the frame

1   As is the case with removal, engine replacement requires considerable care and patience. Replacing the engine necessitates the use of three people and it is important that the machine is standing firmly on level ground. Lift the engine in from the right-hand side of the machine, and manoeuvre the rear of the engine over so that the final driveshaft gaiter can be pulled across to the right and hitched against the output flange. The front of the engine can be moved across so that the output flange and driveshaft flange align..

2   Insert the two engine forward mounting bolts and the nuts which are located by the engine casing. Use a wooden lever between the lower frame tubes to raise the engine the required amount. Fit the single engine rear mounting bolt together with the rider's footrests. Position each rest so that it abuts against the projecting lug and so prevents the footrest from turning. Tighten the engine mounting bolts.

3   Fit the alternator rotor to the end of the crankshaft and tighten the retaining bolt. Replace the alternator cover complete with the stator and field coils which it contains. The two wiring lead connections should be routed under the right-hand edge of

the crankcase base and through between the rear of the crank-case and the frame rear crossmember. Secure the leads by means of the three wiring clips and the guard plate which is retained by the lower front primary drive cover screw. Tighten all the primary drive casing and alternator cover screws evenly.

4   Replace the starter motor in the casing recess so that the starter pinion engages with the intermediate gear. The O ring on the motor boss should be coated with grease to aid installation and improve sealing. Fit and tighten the two motor retaining screws.

5   Reconnect the contact breaker lead block connectors and the two alternator lead connectors with the sockets on the left-hand side of the machine. Track the leads so that they will not become chafed by cycle or engine parts. Connect the main starter motor lead at the starter motor body terminal; the lead is secured by a clip on one rear crankcase bolt. The adjacent crankcase bolt secures the main earth lead from the battery. This too should be reconnected. Reconnect the neutral indicator switch lead and the oil warning pressure switch lead, both of which are secured by a central screw.

6   Replace the four output flange/final drive shaft flange securing bolts, turning the flanges as necessary to gain access. Tighten the bolts evenly and then relocate the drive shaft gaiter.

7   Replace the starter motor cover and retaining bolts and then position the breather hose on the breather cover stub. Do not tighten the screw clip at this stage. Route the clutch cable down-wards and reconnect it with the operating mechanism. The mechanism's rubber cover is a push fit on the spring plate bracket, to the rear of the casing.

8   Place the air cleaner box between the rear frame tubes. Insert the complete carburettor assembly from the left-hand side, reconnecting the throttle cable before final positioning. Fit the carburettors loosely on the inlet stubs. Replace the air-filter mounting bolts, noting that the left-hand rear bolt also secures an earth wire. Install the intake silencer unit on the base of the air-filter box and reconnect the breather tube. Tighten the screw clamps at both ends of the tube. Tighten the screw clips which secure the carburettors to the inlet stubs and the air cleaner box hoses to the carburettor. Fit the two petrol tap vacuum tubes to the take off points on the two outer inlet stubs. The centre stub take off point is fitted with a rubber cap and is not utilised.

9   Install the tachometer drive shaft, securing it by means of the forked plate and screw. Reconnect the tachometer cable and tighten down the knurled ring. Fit the air scoop brackets and then tighten down all the camshaft cover screws. Replace the air scoop.

10 Refit the exhaust system by reversing the dismantling proce-dure. The right-hand silencer bracket may be moved fore and oft within fixed limits to aid refitting, after slackening the pillion footrest bolt and the small domed nut. To aid fitting of the half collars and flanges, use a short length of sellotape to hold the two half collars on each pipe together. Use only new exhaust port gaskets to ensure complete gas tightness.

11 Replace the petrol tank and reconnect the fuel lines and the vacuum lines. Do not omit the small spring clips.

12 Reconnect the spark plug caps and reconnect the battery terminals. Give a final visual check to the electrical connections and replace the two frame side covers. Both are a push fit.

13 Check that the sump drain plug and the middle gearbox drain plug are in place and tight. Install a new filter in the filter chamber and fit the chamber to the base of the crankcase. Ensure that the chamber sealing ring is in good condition. To ensure that oil finds its way to the working surfaces of the engine as soon as possible after initial starting, the oil filter chamber must be primed with oil. To facilitate this operation a filler plug is fitted to the front of the chamber. Using a plastic oil container of the type in which gearbox oil is often supplied, fill the chamber by means of the flexible transparent tube. Continue filling until the oil starts flowing back out of the filler. The correct grade of engine oil must, of course, be used. Allow the oil level to settle and then check the level. If the chamber will accept no more oil, fit the filler plug. Replenish the main engine/

transmission lubrication system by pouring 3.0 litres (6.4/5.3 US/Imp pints) of SAE 20W/50 engine oil into the primary dirve casing. Allow the oil to settle and recheck the level by means of the dipstick in the filler cap. Add more oil if necessary.

14 Replenish the middle gearbox with 360 cc (12/10 US/Imp fl oz) of Hypoid gear oil as follows:

SAE 90 EP	for use above 5°C (41°F)
SAE 80 EP	for use below 5°C (41°F)

Yamaha recommend gearbox oil with a quality and specification rating of GL-4, GL-5, GL-6. This does not apply to the UK where any good quality EP Hypoid oil will be satisfactory. After filling the middle gearbox, check the level by means of the dipstick supplied with the tool kit. The long shank of the dipstick should be used.

50 Starting and running the rebuilt engine

1   Place both petrol taps in the Prime position to allow the unrestricted flow of fuel to the carburettors. As soon as the engine is running revert to the normal 'On' position so that the induction vacuum will control the petrol taps. Start the engine using either the kickstart or the electric starter. Raise the chokes as soon as the engine will run evenly and keep it running at a low speed for a few minutes to allow oil pressure to build up and the oil to circulate. If the red oil pressure indicator lamp is not extinguished, stop the engine immediately and investigate the lack of oil pressure.

2   The engine may tend to smoke through the exhaust initially, due to the amount of oil used when assembling the various components. The excess of oil should gradually burn away as the engine settles down.

3   Check the exterior of the machine for oil leaks or blowing gaskets. Make sure that each gear engages correctly and that all the controls function effectively, particularly the brakes. This is an essential last check before taking the machine on the road.

51 Taking the rebuilt machine on the road

1   Any rebuilt machine will need time to settle down, even if parts have been replaced in their original order. For this reason it is highly advisable to treat the machine gently for the first few miles to ensure oil has circulated throughout the lubrication system and that any new parts fitted have begun to bed down.

2   Even greater care is necessary if the engine has been rebored or if a new crankshaft has been fitted. In the case of a rebore, the engine will have to be run-in again, as if the machine were new. This means greater use of the gearbox and a restraining hand on the throttle until at least 500 miles have been covered. There is no point in keeping to any set speed limit; the main requirement is to keep a light loading on the engine and to gradually work up performance until the 500 mile mark is reached. These recommendations can be lessened to an extent when only a new crankshaft is fitted. Experience is the best guide since it is easy to tell when an engine is running freely.

3   If at any time a lubrication failure is suspected, stop the engine immediately, and investigate the cause. If an engine is run without oil, even for a short period, irreparable engine damage is inevitable.

4   When the engine has cooled down completely after the initial run, recheck the various settings, especially the valve clearances. During the run most of the engine components will have settled into their normal working locations.

49.1 Lift the engine into approximately the correct position

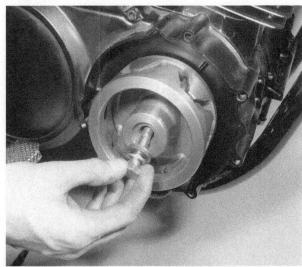

49.3a Fit the alternator rotor and centre bolt

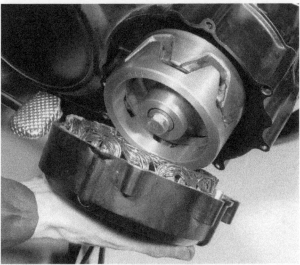

49.3b Replace the alternator cover complete with coils

49.3c Secure the leads below the crankcase

49.5 Connect the main earth lead to the engine casing

49.8 Left-hand rear air box bolt also holds secondary earth lead

49.9a Refit the tachometer drive shaft and ...

49.9b ... reconnect the drive cable

49.10a Use new gaskets at the exhaust ports

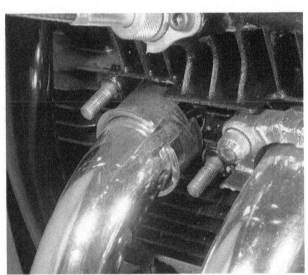

49.10b Apply tape to hold half collars until ...

49.10c ... exhaust flanges have been refitted

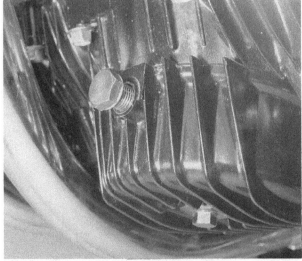

49.13a Remove plug in front of filter chamber

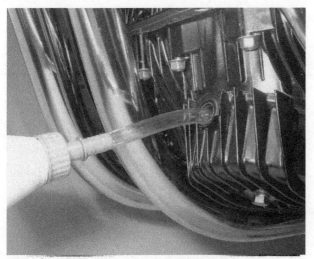

49.13b Fill filter chamber with SAE20W/50 engine oil

49.14 Check level of middle gear case

52 Fault diagnosis: engine

Symptom	Cause	Remedy
Engine will not start	Defective spark plugs	Remove the plugs and lay on cylinder heads. Check whether spark occurs when ignition is switched on and engine rotated.
	Dirty or closed contact breaker points	Check condition of points and whether gap is correct.
	Faulty or disconnected condenser	Check whether points arc when separated. Renew condenser if evidence of arcing.
Engine runs unevenly	Ignition and/or fuel system fault	Check each system independently, as though engine will not start.
	Blowing cylinder head gasket	Leak should be evident from oil leakage where gas escapes.
	Incorrect ignition timing	Check accuracy and if necessary reset.
Lack of power	Fault in fuel system or incorrect ignition timing	See above.
Heavy oil consumption	Cylinder block in need of rebore	Check for bore wear, rebore and fit oversize pistons if required.
	Damaged oil seals	Check engine for oil leaks.
Excessive mechanical noise	Worn cylinder bores (piston slap)	Rebore and fit oversize pistons.
	Worn camshaft drive chain (rattle)	Adjust tensioner or replace chain.
	Worn big-end bearings (knock)	Fit replacement crankshaft assembly.
	Worn main bearings (rumble)	Fit new journal bearings and seals. Renew crankshaft assembly if centre bearings are worn.
Engine overheats and fades	Lubrication failure	Stop engine and check whether internal parts are receiving oil. Check oil level in crankcase.

53 Fault diagnosis - clutch

Symptom	Cause	Remedy
Engine speed increases as shown by tachometer but machine does not respond	Clutch slip	Check clutch adjustment for free play at handlebar lever. Check thickness of inserted plates.
Difficulty in engaging gears. Gear changes jerky and machine creeps forward when clutch is withdrawn. Difficulty in selecting neutral	Clutch drag	Check clutch adjustment for too much free play. Check clutch drums for indentations in slots and clutch plates for burrs on tongues. Dress with file if damage not too great.
Clutch operation stiff	Damaged, trapped or frayed control cable	Check cable and renew if necessary. Make sure cable is lubricated and has no sharp bends.

54 Fault diagnosis - gearbox

Symptom	Cause	Remedy
Difficulty in engaging gears	Selector forks bent Gear clusters not assembled correctly	Renew. Check gear cluster arrangement and position of thrust washers.
Machine jumps out of gear	Worn dogs on ends of gear pinions	Renew worn pinions.
Gearchange lever does not return to original position	Broken return spring	Renew spring.
Kickstarter does not return when engine is turned over or started	Broken or poorly tensioned return spring	Renew spring or retension.

Chapter 2 Fuel system and lubrication

Contents

Specifications

Petrol tank

Capacity	17 litres (4.5/3.8 US/Imp galls)

Carburettors

Make	Mikuni
Type	BS 34, IJ 701
Main jet	145
Jet needle	4H11-3
Needle jet	Y-2
Starter jet	45
Pilot jet	17.5
Float height	26.6 mm
Pilot air screw, no of turns out	2¼
Main air jet	1.0 mm
Pilot air jet	1.6 mm

Oil capacity

Dry	3.5 litres (7.4/6.1 US/Imp pts)
Without filter change	2.8 litres (6/5 US/Imp pts)
With filter change	3.2 litres (6.8/5.6 US/Imp pts)

Oil pump

Type	Trochoid
Outer rotor/housing clearance	0.090 - 0.015 mm (0.0035 - 0.0059 in)
Outer rotor/inner rotor clearance	0.03 - 0.09 mm (0.0011 - 0.0035 in)

1 General description

The fuel system comprises a petrol tank, from which petrol is fed by gravity to the carburettors, via two diaphragm type fuel cocks. There are three positions, ON, RESERVE and PRIMING. Before starting the engine, turn the fuel tap to the ON position this enables the fuel to flow to the carburettors when the engine has started.

If the fuel in the tank is too low to be fed to the carburettors in the ON position, turn the lever to RESERVE position, which provides a small quantity of fuel after the main supply is exhausted. Only when there is no fuel in the carburettors is it necessary to turn to the PRIMING position, which will allow fuel to flow to the carburettors even with the engine stopped. Once the engine has started be sure to return the lever to the ON or reserve position.

For cold starting, a hand operated choke lever attached to the left-hand carburettor is linked to the centre and right-hand instruments, so that the mixture on all three carburettors can be enriched temporarily. Throttle control is effected by the means of a standard cable arrangement which operate a butterfly valve in each carburettor.

Lubrication is by the wet sump principle in which oil is delivered under pressure, from the sump, through a mechanical pump to the working parts of the engine. The pump is of the trochoid type and is driven from a pinion on the left-hand end of the crankshaft via a pinion and shaft integral with the starter motor clutch. Oil is supplied under pressure via a release valve and a full flow oil filter fitted with a paper element. The engine oil supply is also shared by the primary drive and the gearbox.

The middle gear casing which contains the 90° bevel gear, holds its own reservoir of oil, which is remote from that of the engine. The requirement of the bevel gear dictates the use of a high pressure hypoid type gear oil.

2 Petrol tank: removal and replacement

1   The fuel tank is retained at the forward end by two rubber buffers fitted either side of the under side of the tank which fit into cups on the frame top tube. The rear of the tank sits on a small rubber saddle placed across the frame top tube and is retained by a single bolt passing through rubber buffers either side of the tank lug.

2   To remove the tank, pull off the fuel lines at the petrol tap unions where they are held by spring clips. Detach also the vacuum pipes leading to the taps. Raise the seat and remove the rear bolt. Pull the tank up at the rear and back, off the buffers.

3 Petrol taps: removal, filter cleaning and replacement

1   Before either petrol tap is removed for filter cleaning or for attention to the lever, the petrol tank must be drained. Detach the petrol feed pipes and substitute a longer length of suitable piping through which to drain the fuel. Turn the tap to the Prime position to allow the petrol to flow freely.

2   To remove either tap, loosen the two screws which pass through the flange into the petrol tank and withdraw the tap, complete with filter tower. The filter is a push fit on the hollow T piece which projects from the top of the tap.

3   The filter should be cleaned at regular intervals to remove any entrapped matter. Use clean petrol and a soft brush.

4   It is seldom necessary to remove the lever which operates the petrol tap, although occasions may occur when a leakage develops at the joint. Although the tank must be drained before the lever assembly can be removed, there is no need to disturb the body of the tap.

5   To dismantle the lever assembly, remove the two crosshead screws passing through the plate on which the operating positions are inscribed. The plate can then be lifted away, followed by a spring, the lever itself and the seal behind the lever. The seal will have to be renewed if leakage has occurred. Reassemble the tap in the reverse order. Gasket cement or any other sealing medium is NOT necessary to secure a petrol tight seal.

Note that there is an 'O' ring seal between the petrol tap body and the petrol tank, which must be renewed if it is damaged or if petrol leakage has occurred.

2.2a Fuel tank is retained by a single bolt at rear and ...

2.2b ... on two rubber buffers on frame top tube

3.2 Petrol tap is secured by two screws

3.4a Remove the plate and wave washer and ...

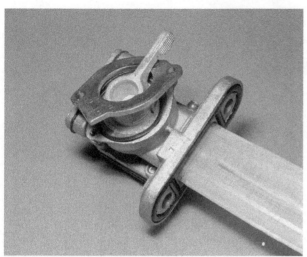

3.4b ... lift out the lever and 'O' ring

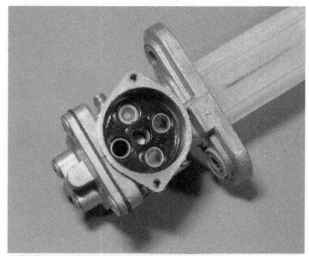

3.4c Seal should be replaced if leakage occurs

4 Petrol feed pipes: examination

1   Synthetic rubber feed pipes are used to convey the flow of petrol from the petrol tap to the float chamber of each of the four carburettors. Each pipe is retained by a wire clip, which must hold the pipe firmly in position. Check periodically to ensure the pipes have not begun to split or crack and that the wire clips have not worn through the surface.

2   Do NOT replace a broken pipe with one of natural rubber, even temporarily. Petrol causes natural rubber to swell very rapidly and disintegrate, with the result that minute particles of rubber would easily pass into the carburettors and cause blockages of the internal passageways. Plastic pipe of the correct bore size can be used as a temporary substitute but it should be replaced with the correct type of tubing as soon as possible since it will not have the same degree of flexibility.

5 Carburettors: removal

1   Detach both right and left-hand frame side covers, both of which are a push fit at their lower edges and retained at the top on two hooks. Disconnect the petrol feed pipes and also the diaphragm vacuum pipes.

2   Remove the two bolts from each side of the frame, which secure the air filter box to the frame. Note that the rearmost left-hand bolt also secures an earth lead. Detach the two intake silencers from below the air filter box. Each is secured by a single screw. Slacken the screw clip which secures the engine breather hose to the air filter box junction.

3   Free the clutch cable from the guide clip which is attached to the left-hand instrument. Loosen the clamps which hold the inlet manifold stubs and the air filter hoses to the carburettors. Move the air filter box rearwards as far as possible so that the air filter hoses leave the carburettor mouths. The carburettors can now be pulled from the stubs as a complete unit and moved out towards the right-hand side of machine. Disconnect the throttle cable from the operating pulley on the throttle control rod by pulling the outer cable up and forwards from the cable abutment and then detaching the nipple from the pulley.

6 Carburettors: dismantling and reassembly

1   In order to dismantle the carburettors they must first be removed from the mounting bracket to which they are each retained by two screws. The carburettors are also connected by brackets, one of which acts as the throttle cable holder. The bracket is retained by two of the four diaphragm cover screws on each carburettor top.

2   Commence by removing the left-hand carburettor from the mounting bracket. Remove the retaining screws as described and then loosen the choke operating claw arm from the operating rod by slackening the grub screw. The carburettor will now pull from position, disconnecting the float chamber interconnection fuel pipes and the throttle linkage. The remaining two carburettors can be removed in a similar manner.

3   Invert each carburettor and remove the four screws that hold the float chamber to its base. Remove the hinge pin that locates the twin float assembly and lift the float from position. This will expose the float needle. The needle is very small and should be put away in a safe place so that it is not misplaced. Make sure that the float chamber gasket is in good condition. Do not disturb the gasket unless leakage has occurred or it appears damaged.

4   Check that the twin floats are in good condition and not punctured. Because they are made of brass it is possible to solder a damaged float. This form of repair should only be made in an emergency, when a set of new floats is not available. Soldering will effect the weight of the float assembly and result in a different petrol level.

5   The needle jet is a push fit in the base of the mixing chamber, being retained by a small 'O' ring. Check the needle jet for wear, together with the jet needle. After lengthy service, these two components should be renewed together, or high petrol consumption will result.

6   The float needle will also wear after lengthy service, and should be closely examined with a magnifying glass. Wear takes the form of a ridge or groove, which will cause the float needle to seat imperfectly. The needle and seating should always be renewed as a pair. The seating is a screw fit in the mixing chamber. Note the O ring and also the tiny filter gauze, which is retained by the seat.

7   The main jet and pilot jet are both housed in the float chamber. The main jet is situated below a plug, which will unscrew from outside the float chamber. Always use a close fitting screwdriver when removing jets, or damage will result.

8   Remove the two remaining screws which retain the carburettor top and lift the top from position, together with the piston spring. Carefully lift the diaphragm from position, bringing with it the piston and jet needle. Carefully check the condition of the diaphragm. If it has developed cracks or holes, it must be renewed as a unit, with the piston. The jet needle is retained by a nylon plate and is a push fit. The jet needle must be renewed

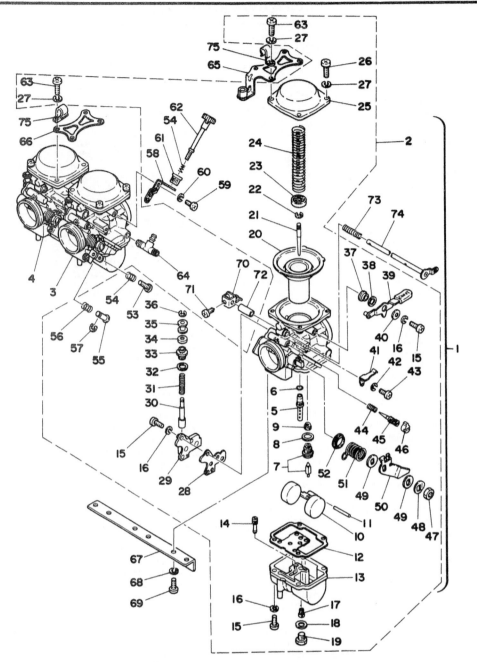

Fig. 2.1. Carburettors - component parts

1	LH carburettor components	18	Sealing washers - 3 off
2	LH carburettor complete	19	Plug - 3 off
3	Centre carburettor complete	20	Piston/diaphragm - 3 off
4	RH carburettor complete	21	Jet needle - 3 off
5	Needle jet (main nozzle) - 3 off	22	'E' clip - 3 off
6	'O' ring - 3 off	23	Needle seat - 3 off
7	Float needle assembly - 3 off	24	Piston spring - 3 off
8	Sealing washer - 3 off	25	Cover - 3 off
9	Filter - 3 off	26	Screw - 4 off
10	Float - 3 off	27	Spring washer - 12 off
11	Pilot pin - 3 off	28	Choke body gasket - 3 off
12	Bowl gasket - 3 off	29	Choke body - 3 off
13	Float bowl - 3 off	30	Choke plunger - 3 off
14	Pilot jet - 3 off	31	Spring - 3 off
15	Screw - 22 off	32	Washer - 3 off
16	Spring washer - 22 off	33	Housing - 3 off
17	Main jet - 3 off	34	Cover - 3 off
		35	Plunger bush - 3 off
		36	'E' clip - 3 off
		37	Pivot bush

38	Washer	57	'E' clip - 2 off
39	Plug - 3 off	58	Bracket
40	Washer	59	Screw
41	Detent spring	60	Spring washer
42	Spring washer	61	Bush
43	Screw	62	Throttle stop main adjuster
44	Spring - 3 off	63	Screw - 8 off
45	Pilot screw - 3 off	64	'T' piece - 2 off
46	Pilot screw cap - 3 off	65	Bracket/cable anchor
47	Nut - 3 off	66	Bracket
48	Washer - 3 off	67	Mounting bar
49	Collar - 4 off	68	Spring washer - 6 off
50	Throttle arm - 3 off	69	Screw - 6 off
51	Throttle return spring - 3 off	70	Choke link arm
52	Washer - 3 off	71	Screw - 3 off
53	Throttle stop screw - 2 off	72	Collar
54 1	Throttle stop spring - 3 off	73	Spring
55	Plunger - 2 off	74	Choke shaft
56	Spring - 2 off	75	Hose guide - 2 off

if worn, as described in paragraph 5.

9 Before the carburettors are reassembled, using the reversed dismantling procedure, each should be cleaned out thoroughly, using compressed air. Avoid using a piece of rag since there is always risk of particles of lint obstructing the internal passageways or the jet orifices.

10 Never use a piece of wire or any pointed metal object to clear a blocked jet. It is only too easy to enlarge the jet under these circumstances and increase the rate of petrol consumption. If compressed air is not available, a blast of air from a tyre pump will usually suffice.

11 Do not use excessive force when reassembling a carburettor because it is easy to shear a jet or some of the smaller screws. Furthermore, the carburettors are cast in a zinc-based alloy, which itself does not have a high tensile strength.

12 When fitting the piston unit note that the diaphragm is provided with a tab which locates in the main body.

13 Do not omit the fuel transfer T pieces when replacing the carburettors on the mounting bracket. If the pipes are not a good push fit they may leak and should therefore be renewed. The interconnecting arm on each throttle valve rod should be fitted between the spring-loaded plunger and the throttle adjusting screw on the interconnecting arm of the next carburettor.

14 After assembly of the carburettors on the mounting bracket, check that all three throttle valves open fully and simultaneously. If differences occur, adjustment should be made using the throttle adjustment screws on the link arms. The remote throttle adjuster screw with the nylon knob, placed between the left-hand and centre carburettors, should be used only to set the throttle opening when adjusting the kickover. **DO NOT** alter the positions of the pilot adjuster screws, each of which is protected by a plastic cap. As stated in Section 8, these screws are set at the factory and must not be tampered with needlessly.

---

7 Carburettors: adjustment for tickover

---

1 Before adjusting the carburettors, a check should be made that when the throttle is fully opened all three butterfly valves are in their fully open position (ie; parallel to the carburettor bore). This check can best be done when the carburettors are off the machine as described in the previous Section.

2 Before adjusting the tickover speed, allow the engine to warm up thoroughly. Tickover speed is regulated by the remote adjuster placed between the left-hand and centre carburettors. Turning the screw inwards increases the speed and vice-versa. The correct idle speed is within the range 1,050 - 1,150 rpm.

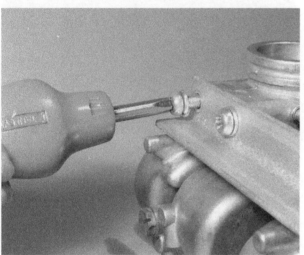

6.1a Carburettors are supported on common mounting bar and ...

6.1b ... interconnected by brackets held by cap screws

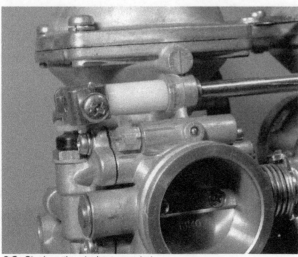

6.2a Slacken the choke control claw

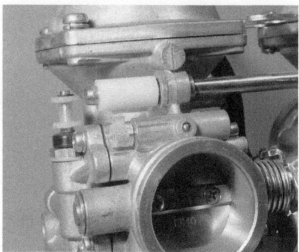

6.2b Note nylon distance piece on link rod

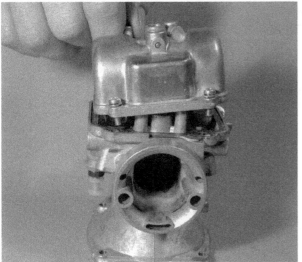

6.3a Carburettor float bowl held by four screws

6.3b Displace pivot and float assembly to ...

6.3c ... remove the float needle

6.5 Needle jet is a push fit in base of body

6.7a Remove the float bowl plug for access to ...

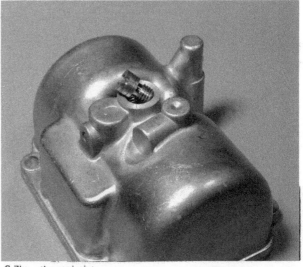

6.7b ... the main jet

6.7c Pilot jet is located inside float bowl

6.8a Remove the piston spring and ...

6.8b ... carefully withdraw the piston/diaphragm

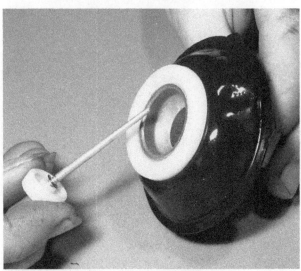

6.8c Displace the jet needle and needle seat

6.12 Note location tab on diaphragm periphery

6.13a Fuel 'T' pieces are a tight push fit

6.13b Link arms slide between plunger and adjuster screw

6.14a Check throttle valves open simultaneously and ...

6.14b ... adjust if necessary, using throttle screw

7.2 Tick-over speed (throttle stop) adjuster knob

8 Carburettors: synchronisation

1   For the best possible performance it is imperative that the carburettors are working in perfect harmony with each other. At any given throttle opening if the carburettors are not synchronised, not only will one cylinder be doing less work but it will also in effect have to be carried by the other cylinders. This effect will reduce the performance considerably.

2   Carburettor synchronisation is governed by the throttle butterfly valves and by the pilot adjuster screws. The valves may be synchronised as described in Section 6.14. Yamaha state that the pilot adjuster screws, which are protected by small plastic anti-tamper caps, are set at the factory using specialised equipment and **should not be touched** during the life of the machine. The adjustments are made accurately not only for performance purposes but in order that the exhaust emissions comply with statutory requirements. It is likely, however, during the normal service life of the machine that the carburettors will need some adjustment due to engine and carburettor wear. As no information is available concerning the correct setting of the carburettors and because vacuum gauges are required to ensure correct synchronisation, it is suggested that the machine be returned to a Yamaha specialist for this work to be carried out.

3   When contemplating resynchronisation of the carburettors, it is well worthwhile first checking that the following clearances and adjustments are correct: Contact breaker gaps, ignition timing, spark plug gaps and if there is any doubt, the valve clearances. What may appear to be a carburettor fault can often be traced to electrical problems.

9 Carburettor settings

1   Some of the carburettor settings, such as the sizes of the needle jets, main jets and needle positions, etc are pre-determined by the manufacturer. Under normal circumstances it is unlikely that these settings will require modification, even though there is provision made. If a change appears necessary, it can often be attributed to a developing engine fault.

2   Apart from alterations of the pilot adjuster screws some alterations to the mid-range mixture characterics may be made by raising or lowering the jet needle. This is accomplished by changing the position of the needle clip. Raising the needle will richen the mixture and lowering the needle will weaken the mixture.

Reference to Chapter 3 will show how the condition of the spark plugs can be interpreted with some experience as a reliable guide to carburettor mixture strength. Flat spots in the carburation can usually be traced to a defective timing advancer. If the advancer action is suspect, it can be detected by checking the ignition timing with a stroboscope.

3   If problems are encountered with fuel overflowing from the float chambers, which cannot be traced to the float/needle assembly, or if consistent fuel starvation is encountered, the fault will probably lie in maladjustment of the float level. It will be necessary to remove the float chamber bowl from each carburettor to check the float level.

If the float level is correct the distance between the uppermost edge of the floats and the flange of the mixing chamber body will be as follows.

26.5 ± 2.5 mm (1.04 ± 0.1 in)

Adjustments are made by bending the float assembly tang (tongue), which engages with the float, in the direction required (see accompanying diagram).

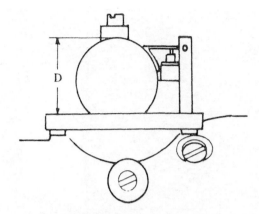

Fig. 2.2. Checking the float level

D = Distance to be measured
26.5 ± 2.5 mm (from gasket surface)

8.2 Anti-tamper plastic cap on pilot screw

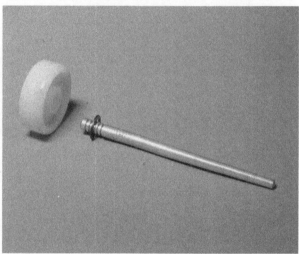

9.2 Mid range mixture altered by moving needle clip

10 Exhaust system

1   Unlike a two-stroke, the exhaust system does not require such frequent attention because the exhaust gases are usually of a less oily nature.

2   Do not run the machine with the exhaust baffles removed, or with a quite different type of silencer fitted. The standard production silencers have been designed to give the best possible performance, whilst subduing the exhaust note to an acceptable level. Although a modified exhaust system or one without baffles, may give the illusion of greater speed as a result of the changed exhaust note, the chances are that performance will have suffered accordingly.

11 Air cleaner: dismantling and cleaning

An air filter cartridge of woven synthetic fibre is fitted in the massive air cleaner box to the rear of the carburettors. Air enters through two special silencer ducts at the base of the air cleaner box and is drawn up through a plenum chamber and through the air filter element to the carburettor. A rubber pipe runs from the breather cover on the crankcase to a union on the lower front edge of the filter box. This is fitted so that all carbon based vapours expelled from the engine are recirculated through the carburettors and then burnt in the cylinders.

2   To gain access to the filter element, raise the dualseat and remove the two screws which secure the cleaner box cover. Lift the cover away and withdraw the element and element holder. Detach the holder plate by unscrewing the thumb nut.

To clean the element, tap it lightly to loosen the accumulation of dust and then use a soft brush to sweep the dust away. Alternatively, compressed air can be blown into the element from the inside. Remember the element is composed of woven fabric and is easily damaged if handled roughly.

If the element is damp or oily it must be renewed. A damp or oily element will have a restrictive effect on the breathing of the carburettor and will almost certainly affect the engine performance.

On no account run without the air cleaners attached, or with the element missing. The jetting of the carburettors takes into account the presence of the air cleaners and engine performance will be seriously affected if this balance is upset.

To replace the element, reverse the dismantling procedure. Give a visual check to ensure that the inlet hoses are correctly located and not kinked, split or otherwise damaged. Check that the air cleaner boxes are free from splits or cracks.

11.2 Air filter element is a slide fit in air box

12 Engine lubrication

1   The engine lubrication system, which is also shared by the gearbox and primary drive, is of the wet sump type, where the oil reservoir is contained in the sump. The sump has a capacity of 3.5 litres (7.4/6.1 US/Imp pints) including the oil filter which holds approximately 0.4 litres (0.8/0.7 US/Imp pints).

2   A trochoid oil pump driven from a pinion on the left-hand end of the crankshaft, via a pinion and shaft integral with the starter motor clutch, delivers oil from the sump to the rest of the engine. The pump components are retained in a casting which also houses a pressure release valve and on all but early models, a non-return valve. Oil is picked up from the sump through a wire gauze strainer, which protects the oil pump from any large particles of foreign matter. The delivery section of the pump feeds oil at a preset pressure via a pressure release valve, which by-passes oil to the sump if the pressure exceeds the preset limit. As a result, it is possible to maintain a constant pressure in the lubrication system. The standard pressure is 42 psi (3 kg/cm^2) when the engine is running at cruising rpm and the oil is at normal running temperature.

3   Since the oil flow will not, under normal circumstances, actuate the pressure release valve, it passes directly through the full flow filter which has a replaceable element, to filter out any impuritites which may otherwise pass to the working parts of the engine. The oil filter unit has its own by-pass valve to prevent the cut-off of the oil supply if the filter element has become clogged.

4   Oil from the filter is fed direct to the crankshaft and big-end bearings with a separate pressure feed to the camshaft and valve assemblies. Oil is also fed directly to the mainshaft and layshaft in the gearbox and thence to the pinion bushes. Surplus oil drains to the sump where it is picked up again and the cycle is repeated.

5   An oil pressure warning light is included in the lubrication circuit to give visual warning by means of an indicator light, if the pressure should fall to a low level.

13 Oil pump: dismantling, examination and reassembly

1   The oil pump can be removed from the engine while the engine is still in the frame. However, it will be necessary to detach the sump after draining the engine oil.

2   The oil pump is retained by three screws and is located on two hollow dowels. On all but early models, one screw is obscured by the filter screen funnel mouth, which must be detached to gain access. The screen is retained by three screws and the cast

funnel by two screws.

3   After removal of the pump, unscrew the three housing cover screws and separate the cover, complete with drive pinion and shaft, from the main body. Note the small drive pin which is a light push fit in the end of the driveshaft. Lift out both the inner and outer pump rotors.

4   Since the oil pump must be cleaned, both the pressure release valve and the non-return valve should be removed from the pump body. Each is retained together with its tension spring by a backing washer and circlip. Using a round nosed tool, depress the spring and washer whilst removing the securing circlip. Release the pressure and remove the washer, spring and plunger.

5   Wash all the pump components with petrol and allow them to dry before carrying out an examination. Before partially reassembling the pump for various measurements to be carried out, check the casting for breakage or fracture, or scoring on the inside perimeter.

6   Reassemble the pump rotors and measure the clearance between the outer rotor and the pump body, using a feeler gauge. If the measurement exceeds the service limit of 0.090 mm (0.0035 in), the rotor or the body must be renewed, whichever is worn. Measure the clearance between the outer rotor and the inner rotor, using a feeler gauge. If the clearance exceeds 0.090 mm (0.0035 in) the rotors must be renewed as a set. It should be noted that one face of both the inner and outer rotor is punch marked. The punch marks should face towards the main pump casing during measurements and on reassembly.

7   Examine the rotors and the pump body for signs of scoring, chipping or other surface damage which will occur if metallic particles find their way into the oil pump assembly. Renewal of the affected parts is the only remedy under these circumstances, bearing in mind that the rotors must always be replaced as a matched pair.

Reassemble the pump components by reversing the dismantling procedure. Remember that the punch marked faces of the rotors must face towards the main pump body. The component parts must be ABSOLUTELY clean or damage to the pump will result. Replace the rotors and lubricate them thoroughly before refitting the cover.

8   Before refitting the pump, it should be filled with oil by holding it inlet side upwards and pouring clean engine oil in through the inlet orifice, whilst simultaneously rotating the drive pinion in an anticlockwise direction. When oil can be seen to escape from the feed aperture, the pump is sufficiently primed for replacement. Check that the sealing O ring is in good condition, to prevent oil pressure loss at the mating surfaces.

9   If damage occurs to the pump pinion, the latter may be removed after loosening and detaching the centre nut. The pinion is located by a cross-pin which is a push fit in the shaft and provides the drive medium.

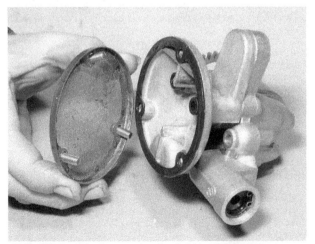

13.2a Filter mesh screen and gasket held by three screws

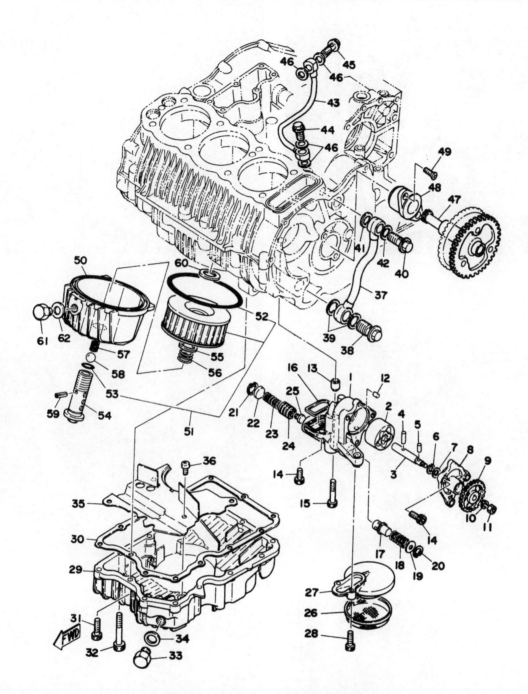

Fig. 2.3. Lubrication system - component parts

1 Oil pump body	17 Pressure relief plunger	33 Drain plug	49 Countersunk screw - 2 off
2 Rotor assembly	18 Spring	34 Sealing washer	50 Oil filter chamber
3 Drive shaft	19 Washer	35 Baffle plate	51 Oil filter assembly
4 Drive pin (rotor)	20 Circlip	36 Screw - 4 off	52 'O' ring
5 Drive pin (pinion)	21 Circlip	37 Oil feed pipe	53 'O' ring
6 Circlip	22 Plug	38 Banjo bolt	54 Filter bolt
7 Washer	23 Spring	39 Sealing washer - 2 off	55 Washer
8 Pump cover	24 Non-return valve plunger	40 Banjo bolt	56 Spring
9 Drive pinion	25 Rubber valve seat	41 Sealing washer	57 Spring
10 Spring washer	26 Oil strainer	42 Sealing washer	58 Pressure release ball
11 Nut	27 Strainer funnel	43 Camshaft feed pipe	(½ inch)
12 Dowel pin	28 Socket screw - 2 off	44 Banjo bolt	59 Roll pin
13 Hollow dowel - 2 off	29 Sump	45 Banjo bolt	60 Washer
14 Socket screw - 4 off	30 Gasket	46 Sealing washer - 4 off	61 Filler plug
15 Socket bolt - 2 off	31 Socket screw - 2 off	47 Oil pump driven shaft	62 Sealing washer
16 Seal	32 Socket bolt	48 Shaft bush	

13.2b Remove filter funnel to allow ...

13.2c ... removal of the oil pump

13.3a Pump housing cover is held by three screws

13.3b Drive pin locates with inner rotor

13.4a Oil pressure relief valve components

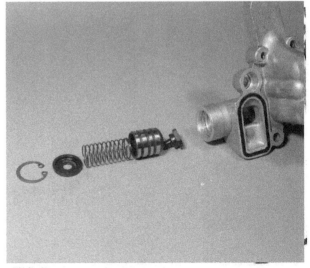

13.4b Non-return valve components

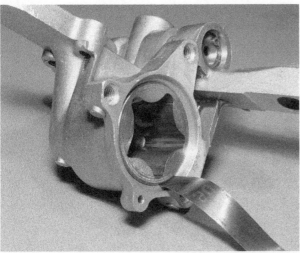
13.6a Check outer rotor/housing clearance and ...

13.6b ... inner rotor/outer rotor clearance

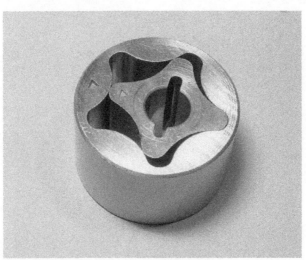
13.6c Note punch marks on inner face of rotors

14 Oil filter: renewing the element

1   The oil filter is contained within a detachable chamber fitted to the base of the sump. Access to the element is made by unscrewing the filter chamber centre bolt which will bring with it the chamber and also the element. Before removing the chamber place a receptacle beneath the engine to catch the engine oil contained in the filter chamber.

2   When renewing the filter element it it wise to renew the filter cover O ring at the same time. This will obviate the possibility of any oil leaks. Do not overtighten the centre bolt on replacement; the correct torque setting is 3.0 - 3.4 kg m (22 - 25 ft lbs).

3   The filter by-pass valve, comprising a plunger and spring, is situated in the bore of the filter cover centre bolt. It is recommended that the by-pass valve be checked for free movement during every filter change. The spring and plunger are retained by a pin across the centre bolt. Knocking the pin out will allow the spring and plunger to be removed for cleaning.

4   Never run the engine without the filter element or increase the period between the recommended oil changes or oil filter changes.

Engine oil should be changed every 3,000 miles and the element changed every 6,000 miles. Use only the recommended viscosity.

15 Oil pressure warning lamp

1   An oil pressure warning lamp is incorporated in the lubrication system to give immediate warning of excessively low oil pressure.

2   The oil pressure switch is screwed into the crankcase, below and to the rear of the alternator casing. The switch is interconnected to a warning light on the lighting panel on the handlebars. The light should be on whenever the ignition is on but will usually go out at about 1,500 rpm.

3   If the oil warning lamp comes on whilst the machine is being ridden the engine should be switched off immediately, otherwise there is a risk of severe engine damage due to lubrication failure. The fault must be located and rectified before the engine is restarted and run, even for a brief moment. Machines fitted with plain shell bearings rely on high oil pressure to maintain a thin oil film between the bearing surfaces. Failure of the oil pressure will cause the working surfaces to come into direct contact, causing overheating and eventual seizure.

16 Fault diagnosis: fuel system and lubrication

Symptom	Cause	Remedy
Engine gradually fades and stops	Fuel starvation	Check vent hole in filler cap. Sediment in float chamber. Dismantle and clean.
Engine runs badly. Black smoke from exhausts	Carburettor flooding	Dismantle and clean carburettor. Check for punctured float or sticking float needle.
Engine lacks response and overheats	Weak mixture Air cleaner disconnected or hose split Modified silencer has upset carburation	Check for partial block in carburettors. Reconnect or renew hose. Replace with original design.
Oil pressure warning light comes on	Lubrication system failure	Stop engine immediately. Trace and rectify fault before re-starting.
Engine gets noisy	Failure to change engine oil when recommended	Drain off old oil and refill with new oil of correct grade. Renew oil filter element.

Chapter 3 Ignition system

Contents

Specifications

Alternator

Make	Hitachi
Type	Ld 120 - 02
Output	280 watts
No load voltage	14.5 $\pm$ 0.5 V
Field coil resistance (at 20°C)	4.04 $\pm$ 0.4 ohms
Stator coil resistance (at 20°C)	0.48 $\pm$ 0.05 ohms

Contact breaker gap

Dwell angle	0.3 - 0.4 mm (0.012 - 0.016 in)
Ignition timing:	
Retarded	10° @ 1100 rpm
Advanced	38.5° @ 2,900 rpm, advance starts 1550 $^{+200}_{-0}$ rpm

Ignition coil

Make	Hitachi
Type	CM11 - 52A
Primary resistance (at 20°C)	4.0 $\pm$ 0.4 ohms
Secondary resistance (at 20°C)	11.0 $\pm$ 1.1 K ohms

Condenser

Capacity	0.22 microfarad $\pm$ 10%

Spark plugs

Make	*NGK or Motorcraft
Type	BP - 7ES or AG2
Gap	0.7 - 0.8 mm (0.028 - 0.032 in)

manufacturer's recommendation.

1 General description

1 The spark necessary to ignite the petrol vapour in the combustion chambers is supplied by a battery and three ignition coils (one for each cylinder). There are three contact breakers with a single lobe cam that determines the exact moment at which the spark will occur in each cylinder. When the contact points separate, the low tension circuit is broken and a high tension voltage is developed by the coil which jumps the air gap across the points of the spark plug due to fire, and ignites the mixture.

2 The AC generator attached to the right-hand side of the crankshaft generates an alternating current which is rectified and used to change the 12 volt battery. The rectifier is located to the rear of the air cleaner box below the battery mounting box. The three ignition coils are mounted in the frame cradle under the petrol tank. The contact breaker assembly is located at the front of the engine on the left-hand side, under a cover held on by three screws. The whole circuit is operated by a switch actuated by a key, mounted on a panel in between the speedometer and tachometer.

2 Crankshaft alternator: checking the output

If the performance of the alternator is suspect, the output should be checked, using a multimeter, as described in Chapter 6, Section 2.

3 Ignition coils: checking

1 Each cylinder has its own ignition circuit and if one cylinder only misfires, the other two ignition circuits may be disregarded. The components most likely to fail in the circuit that is defective are the condenser and the ignition coil since contact breaker faults should be obvious on close examination. Replacement of the existing condenser will show whether the condenser is at fault, leaving by the process of elimination the ignition coil.

2 The ignition coil can best be checked using a multimeter set to the resistance position. Detach the orange lead and red/white lead at their snap connectors and detach the spark plug cap from the spark plug. Measure the primary winding resistance and the secondary winding resistance by connecting the multimeter as shown in the accompanying diagram.

The resistance values for each circuit should be as follows:

Primary coil resistance   4.0 ohms ± 10% at 20°C
Secondary coil resistance   11.0 K ohms ± 10% at 20°C

Slight variation may be encountered if the ambient temperature departs greatly from that given. If the values differ from those given, the coil is faulty.

3 If a multimeter is not available, and by means of testing, the other components have been found to be satisfactory, the following method may be used to give an estimation of the coil's condition. Remove the suppressor cap and bare the inner wire. Remove the contact breaker cover and turn the engine over until the contact breaker points relevant to the coil to be tested are closed. Turn the ignition on and using an insulated screwdriver flick the points open and shut. If the bared end of the HT lead is held approximately 5 mm from an earthing point (the cylinder head) whilst this is done, a blue spark should jump the gap. If the spark is unable to jump a gap, or is yellowish in colour, the coil is probably at fault.

3 The ignition coils are sealed units and it is not possible to effect a satisfactory repair in the event of failure. A new coil must be fitted.

4 The ignition coils are mounted underneath the petrol tank. They bolt direct to a metal plate across the triplex top frame tubes and face in a rearward direction, parallel to the axis of the machine.

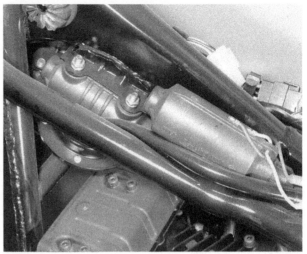

3.4 Ignition coils are mounted below petrol tank

4 Contact breaker adjustments

1 To gain access to the contact breaker assembly it is necessary to remove the three screws holding the cover plate on the right-hand front of the crankcase, and the plate itself.

2 Rotate the engine by turning it over slowly with the kick start until one set of points is fully open. Examine the faces of the contacts for pitting or burning. If badly pitted or burnt they should be removed and treated as described in Section 5 of this Chapter.

3 Adjustment is carried out by slackening the screw on the base of the fixed contact and adjusting the gap 0.3 to 0.4 mm (0.012 - 0.016 in) by moving the stationary contact with a screwdriver. Retighten the screw after adjustment is made and repeat the same operation on the other two sets of contacts. Do not forget to double check after you have tightened the setting screws, in case the setting has altered.

4 Before replacing the cover and gasket, place a slight smear of grease on the cam and a few drops of oil on the felt pads. Do not over lubricate for fear of oil getting on to the points, so that poor electrical contact results.

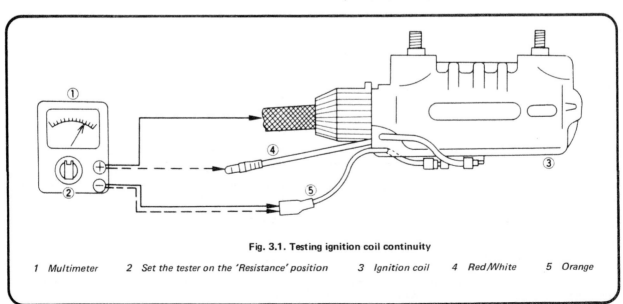

Fig. 3.1. Testing ignition coil continuity

1 Multimeter   2 Set the tester on the 'Resistance' position   3 Ignition coil   4 Red/White   5 Orange

5 Contact breaker points: removal, renovation and replacement

1   If the contact breaker points are burned, pitted or badly worn, they should be removed for dressing. If it is necessary to remove a substantial amount of material before the faces can be restored, the points should be renewed.

2   To remove the contact breaker points, detach the circlip which secures the moving contact to the pin on which it pivots. Remove the nut and bolt which holds the low tension lead to the contact point return spring and spring anchor. Carefully note the position of the insulating washers, which are fitted to the nut and bolt, so that they can be correctly replaced. Lift the moving contact off the pivot pin.

3   The fixed contact assembly is held to the base plate by a single crosshead screw and washer.

4   The points should be dressed with an oil stone or fine emery cloth. Keep them absolutely square throughout the dressing operation, otherwise they will make angular contact on reassembly, and rapidly burn away. If emery cloth is used, it should be backed by a flat strip of steel. This will reduce the risk of rounding the edges of the points.

5   Replace the contacts by reversing the dismantling procedure, making quite certain that the insulating washers are fitted in the correct way. In order for the ignition system to function at all, the moving contact and the low-tension lead must be perfectly

insulated from the base plate and fixed contact. It is advantageous to apply a very light smear of grease to the pivot pin, prior to replacement of the moving contact.

6   Check, and if necessary, re-adjust the contact breaker points when they are in the fully open position.

6 Condenser: removal and replacement

1   There are three condensers contained in the ignition system, wired in parallel with the points. If a fault develops, ignition failure is liable to occur.

2   If the engine proves difficult to start, or misfiring occurs, it is possible a condenser is at fault. To check, separate the contact points by hand when the ignition is switched on. If a spark occurs across the points as they are separated by hand and they have a blackened and a burnt appearance, the condenser associated with that set of points can be regarded as unserviceable.

3   To test a condenser, sophisticated test equipment, is necessary. In view of the small cost involved it is preferable to fit a new one and observe the effect on engine performance, by substitution.

4   The three condensers are clamped to the contact breaker back plate and are wired individually to each set of points. The screws that hold the condensers to the back plate also form the earth connection and should always be checked for tightness.

4.3 Slacken this screw to adjust points gap

4.4 Lubricate the cam feed wicks sparingly

5.2a Note sequence of insulating washers

5.2b Circlip retains moving contact to base pivot

6.4 Each condenser is held by a single screw

7 Ignition timing: checking and resetting

1   The timing can be checked both manually with the engine stationary, and with the engine running, using a stroboscope lamp. In both cases the operation should be carried out by referring to the timing marks stamped on the automatic timing unit (ATU) and the index pointer, which is secured to the casing behind the contact breaker assemblies baseplate. The ATU is stamped with the following marks:

1F	LH cylinder: retarded firing point
2F	Centre cylinder: retarded firing point
3F	RH cylinder: retarded firing point
T	TDC mark for each cylinder

In addition, there are three pairs of scribed parallel lines which indicate the full advance position for each cylinder. These lines are unmarked.

2   Whichever method of timing is chosen, the LH cylinder must be checked first and if necessary adjusted. To check the ignition timing manually a small light bulb should be connected between the low tension terminal on the contact breaker unit in question, and a suitable earth point on the engine. The light will remain on when the points are open and extinguish when the points close, if the ignition is switched on. A multimeter connected in similar fashion, with the resistance range selected, may be substituted in place of the light. Commence by connecting the light to the left-hand contact breaker. Switch the ignition on and using the hexagon provided on the end of the contact breaker camshaft, turn the engine in an anti-clockwise direction. The light should come on the moment the 1F mark aligns with the index pointer in the crankcase. If the ignition timing is incorrect, slacken the three screws which pass through the contact breaker assemblies baseplate. Rotate the plate in the direction desired and then tighten the screws. Recheck that the ignition timing is now correct.

3   The timing on each of the other cylinders may be checked, using the same procedure. The top contact breaker controls the centre cylinder, for which the timing mark is 2F. If the timing is incorrect **do not** move the baseplate, instead the mounting plate on which the relevant contact breaker unit is mounted should be moved. The plate is held by two screws.

4   To check the timing using a stroboscope, interconnect the timing light with the LH cylinder in the manner directed by the lamp's manfuacturer. Start the engine and allow it to idle. Point the timing lamp at the ATU through the aperture in the contact breaker main baseplate and check that the 1F mark aligns with the index mark on the casing. If alignment is not perfect, stop the engine and adjust the baseplate as described for manual adjustment. With the adjustment correct, increase the speed of the engine to above 3000 rpm. The index pointer should align with the small space between the two parallel advance lines for this cylinder. If the retarded ignition timing is correct, but the advance timing is incorrect, the condition of the ATU should be checked as described in the next Section. The advance commences at approximately 1600 rpm. When the timing is correct for No. 1 cylinder (LH) check and adjust the timing for the centre and right-hand cylinders.

5   The index pointer in the casing is retained by a single screw passing through an elongated hole. The elongated hole was provided to allow precise adjustment of the pointer position during original assembly. **NEVER** move the pointer, as the correct datum for ignition timing will then be lost. If the pointer is moved accidentally, it must be repositioned using a degree disc attached to the crankshaft. The pointer must be located so that when the LH piston is at 10° BTDC, the index mark aligns exactly with the 1F mark.

7.2 LH cylinder timing adjustment screws

7.3 A = Centre cylinder timing adjuster screws; B = RH cylinder screws

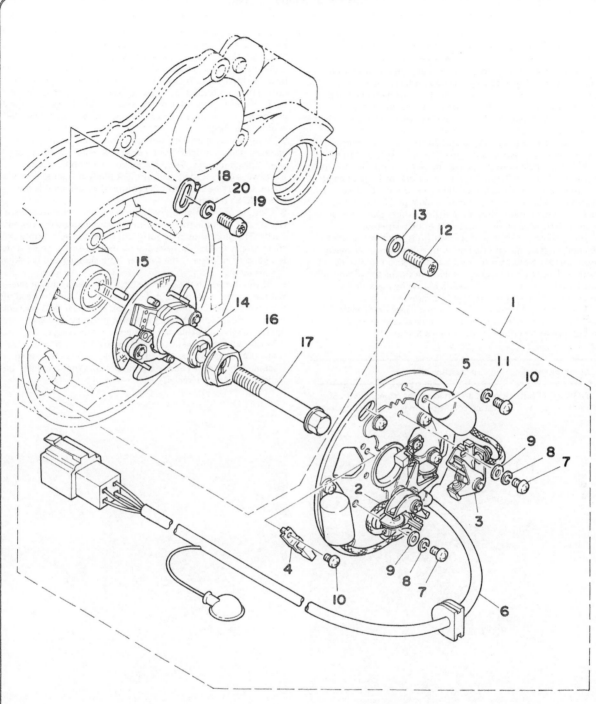

Fig. 3.2. Contact breaker and ATU - component parts

1	Contact breaker assembly complete	11	Spring washer - 3 off
2	LH contact breaker unit	12	Screw - 3 off
3	RH contact breaker unit	13	Plain washer - 3 off
4	Lubrication wick - 2 off	14	Automatic timing unit (ATU)
5	Condenser - 3 off	15	Drive pin
6	Lead	16	Engine turning hexagon
7	Screw - 3 off	17	Bolt
8	Spring washer - 3 off	18	Timing index pointer
9	Plain washer - 3 off	19	Screw
10	Screw - 5 off	20	Spring washer

8 Automatic timing unit: examination

1 The automatic timing mechanism rarely requires attention, although it is advisable to examine it periodically, when the contact breaker is receiving attention. It is retained by a small bolt and washer through the centre of the integral contact breaker cam and can be pulled off the end of the camshaft when the contact breaker plate is removed.

2 The unit comprises spring loaded balance weights, which move outward against the spring tension as centrifugal force increases. The balance weights must move freely on their pivots and be rust-free. The tension springs must also be in good condition. Keep the pivots lubricated and make sure the balance weights move easily, without binding. Most problems arise as a result of condensation, within the engine, which causes the unit to rust and balance weight movement to be restricted.

3 The automatic timing unit mechanism is fixed in relation to the crankshaft by means of a dowel. In consequence the mechanism cannot be replaced in anything other than the correct position. This ensures accuracy of ignition timing to within close limits, although a check should always be made when reassembly of the contact breaker is complete.

4 The correct functioning of the auto-advance unit can be checked when carrying out ignition timing checks using a stroboscope as described in the previous Section.

9 Spark plugs: checking and resetting the gaps

NGK BP-7 ES spark plugs are fitted to the Yamaha XS 750 as standard. Certain operating conditions may indicate a change in spark plug grade, but generally the type recommended by the manufacturer gives the best all round service.

2 Check the gap of the plug points every three monthly or 2,000 mile service. To reset the gap, bend the outer electrode to bring it closer to, or further away from the central electrode until a 0.7 mm (0.028 in) feeler gauge can be inserted. Never bend the centre electrode or the insulator will crack, causing engine damage if the particles fall into the cylinder whilst the engine is running.

3 With some experience, the condition of the spark plug electrodes and insulator can be used as a reliable guide to engine operating conditions. See the accompanying photographs.

4 Always carry a spare pair of spark plugs of the recommended grade. In the rare event of plug failure, they will enable the engine to be restarted.

5 Beware of over-tightening the spark plugs, otherwise there is risk of stripping the threads from the aluminium alloy cylinder heads. The plugs should be sufficiently tight to seat firmly on their copper sealing washers, and no more. Use a spanner which is a good fit to prevent the spanner from slipping and breaking the insulator.

6 If the threads in the cylinder head strip as a result of over-tightening the spark plugs, it is possible to reclaim the head by the use of a Helicoil thread insert. This is a cheap and convenient method of replacing the threads; most motorcycle dealers operate a service of this nature at an economic price.

7 Make sure the plug insulating caps are a good fit and have their rubber seals. They should also be kept clean to prevent tracking. These caps contain the suppressors that eliminate both radio and TV interference.

8.1 Timing marks on ATU

8.2 Check balance weight pivots and springs

Electrode gap check – use a wire type gauge for best results.

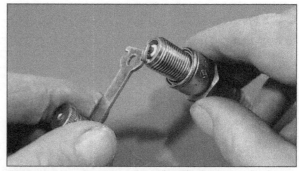

Electrode gap adjustment – bend the side electrode using the correct tool.

Normal condition – A brown, tan or grey firing end indicates that the engine is in good condition and that the plug type is correct.

Ash deposits – Light brown deposits encrusted on the electrodes and insulator, leading to misfire and hesitation. Caused by excessive amounts of oil in the combustion chamber or poor quality fuel/oil.

Carbon fouling – Dry, black sooty deposits leading to misfire and weak spark. Caused by an over-rich fuel/air mixture, faulty choke operation or blocked air filter.

Oil fouling – Wet oily deposits leading to misfire and weak spark. Caused by oil leakage past piston rings or valve guides (4-stroke engine), or excess lubricant (2-stroke engine).

Overheating – A blistered white insulator and glazed electrodes. Caused by ignition system fault, incorrect fuel, or cooling system fault.

Worn plug – Worn electrodes will cause poor starting in damp or cold conditions and will also waste fuel.

10 Fault diagnosis: Ignition system

Symptom	Cause	Remedy
Engine will not start	Faulty ignition switch	Operate switch several times in case contacts are dirty. If lights and other electric function, switch may need renewal.
	Starter motor not working	Discharged battery. Use kickstart until battery is recharged.
	Short circuit in wiring	Check whether fuse is intact. Eliminate fault before switching on again.
	Completely discharged battery	If lights do not work, remove battery and recharge.
Engine misfires	Faulty condenser in ignition circuit	Renew condenser and re-test.
	Fouled spark plug	Renew plug and have original cleaned.
	Poor spark due to generator failure and discharged battery	Check output from generator. Remove and recharge battery.
Engine lacks power and overheats	Retarded ignition timing	Check timing and also contact breaker gap. Check whether auto-timing unit has jammed.
Engine 'fades' when under load	Pre-ignition	Check grade of plugs fitted; use recommended grades only.

Chapter 4 Frame, forks and final drive

Contents

Specifications

Front forks
Oil capacity (per leg)	170 cc (5.75/6.0 US/Imp fl oz)
Oil (damping fluid) type	SAE 20W ATF or fork oil

Front fork travel
750D	140 mm (5.5 in)
750-2D	167 mm (6.6 in)

Rear suspension
Type	Swinging arm on 5-way adjustable suspension units.
Travel	76.2 mm (3.0 in)

1 General description

The frame used on the Yamaha XS 750 is of the full cradle type, in which the engine/gearbox unit is supported by duplex tubes running below the crankcase. The upper part of the frame comprises a single large section top tube, running from the steering head lug to a point forward of the dualseat, which is flanked by two tubes extended to form a support for the dual-seat and upper mounting points for the rear suspension units. Lugs for a centre stand are fitted below the frame, the side stand being fitted to a large fabricated bracket bolted to the left-hand cradle tube.

The front forks are of the conventional telescopic type having two way damping controlled by fluid. The forks fitted to the 750 - 2D model differ slightly in characteristics from those of the 750D, the later type having marginally lower spring and damping rates and increased travel. Rear suspension is provided by a swinging arm fork, pivoting on two tapered roller bearings located by adjustable stubs, and is supported by two fluid filled suspension units adjustable to five positions to give differing spring rates. The left-hand longitudinal swinging arm member also serves as a torque tube through which the final driveshaft passes, and to which is attached the final drive gear casing.

2 Front forks: removal from the frame

1 It is unlikely that the front forks will need to be removed from the frame as a complete unit unless the steering head bearings require attention or the forks are damaged in an accident.
2 Start by removing either the control cables from the handlebar control levers or the levers, complete with cables. The shape of the handlebars and the length of the control cables will probably dictate which method is used. Ensure that the four screws retaining the front brake master cylinder cover are tight before detaching the complete brake lever/master cylinder assembly from the handlebars. The lever assembly is retained by a clamp and two bolts. Tie the assembly to a suitable part of the frame that is not going to be disturbed so that the weight is not taken by the brake hose. It is quite possible to remove the forks without having to separate the brake components from one another. As this method eliminates the necessity of bleeding the front brake on reassembly, it is herein described.
3 Detach the handlebars from the top (crown) yoke. They are retained by two clamps, each of which is secured by two bolts. Remove the one or two screws which pass through the lower edge of the headlamp rim and which secure the rim in place and pull the complete rim/lens assembly away at the lower edge. Pull

2.4 Disconnect instrument drive cables at heads and ...

2.6 ... detach speedometer cable at front wheel

2.9a Front brake hoses interconnect at junction

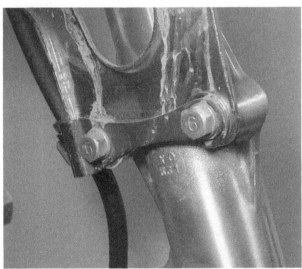

2.9b Mudguard is held by two bolts in each leg

2.11a Loosen upper yoke pinch bolt and ...

2.11b ... the two bolts in lower yoke

the wiring socket from the back of the reflector unit. Lift the dualseat and disconnect the negative lead (−) from the battery terminal followed by the positive (+) lead. This will isolate the electrical system and prevent short circuits occuring when the headlamp and instruments are detached. The headlamp shell need not be separated from the fork shrouds at this stage. The complete assembly together with the front indicator lamps can be removed as a unit after the upper yoke has been lifted away. Remove the bolt which controls the vertical beam adjustment. The bolt passes through a bracket on the shell and a bracket on the lower yoke.

4   Disconnect the speedometer and tachometer drive cables; each is secured by a knurled ring. The instrument heads and warning light panel are retained on a common mounting bracket, secured to the upper yoke by two bolts. Remove the bolts and lift the complete assembly away, after making any necessary disconnection of the wiring leads.

5   If the machine is not already resting on the centre stand, support it in this manner on firm, level ground. Balance the machine so that the front wheel is clear of the ground and place some packing under the crankcase so that if the machine should inadvertently tip forward, it will not roll off the centre stand.

6   Remove the speedometer drive cable from the drive gearbox positioned to the left of the front wheel hub, on the wheel spindle. The cable is retained by a knurled ring.

7   To permit removal of the front wheel, one caliper unit must be detached from the fork leg to which it is secured. In this case, however, both units should be detached, to allow fork leg removal. Each caliper is retained by two bolts which thread into the caliper main body.

8   The front wheel can now be released by withdrawing the spindle, which passes through the left-hand fork leg and is retained by a castellated nut, and split pin. Note that it will be necessary to slacken the two nuts which secure the clamp around the head of the spindle, at the extreme end of the left-hand fork leg. The head of the spindle is drilled, so that a tommy bar can be inserted, to aid removal. Place one foot below the wheel to support the weight as the spindle is withdrawn, and lower the wheel down and out, towards the front . Mark the tyre so that the wheel is refitted the samy way round.

9   The two lower brake hoses are interconnected at a junction piece, retained on the lower yoke by single bolt. Free the junction at this stage so that when the mudguard is detached, the two calipers, complete with the hoses, junction piece and master cylinder, can be lifted away from the machine as a unit. Remove the two bolts which pass into each fork leg and secure the mudguard bracket and each hydraulic hose clamp. Lift the mudguard out carefully from between the legs.

10   Unscrew the crown bolt from the centre of the upper yoke and remove it, together with the heavy washer. Slacken the pinch bolt at the rear of the upper yoke.

11   Loosen the pinch bolts which clamp the yokes to the fork legs The fork legs may now be pulled downwards individually, out of position in the yokes. If required, a rawhide mallet and block of wood can be used to aid removal.

12   Using a rawhide mallet carefully, ease the upper yoke from position on the steering stem. Support the fork shroud/headlamp shell assembly and then remove it as soon as it is freed.

13   Using a C spanner, remove the locking peg nut from the top of the steering stem. Support the lower yoke with one hand and then remove the second peg nut. The lower yoke, complete with the steering stem, can now be lowered from position. The lower tapered bearing inner race will remain on the steering stem leaving the outer race in position. The complete upper tapered roller bearing will remain in the top of the head lug casting.

3   Front forks: dismantling

1   The fork legs can be dismantled individually, without need to disturb the steering head, steering head bearings, the headlamp and instrument assembly. The preliminary dismantling is accomplished by following the procedure detailed in paragraphs 5-8, 11 and parts of paragraph 9 of the preceding Section. Remove the small drain screws and allow the damping oil to drain, before dismantling. This may be facilitated by pumping each fork leg up and down to expel the fluid.

2   If both fork legs are to be dismantled, strip them separately using an identical procedure. There is less chance of unwittingly interchanging parts if this approach is adopted.

3   Using an hexagonal Allen key, remove the socket screw which is recessed into the curved portion of the lower fork leg, through which the wheel spindle normally passes. The socket screw retains the damper rod in position inside the fork leg.

4   Using a piece of inner tubing to protect the chrome finish, grip the stanchion between the jaws of a vice. The fork spring is retained by a plug, which itself is secured by a circlip. To remove the circlip, the plug must be pressed inwards slightly against the pressure of the spring. It is advisable to have an assistant available to help carry out this operation. Remove the rubber bung from the top of the stanchion, to gain access to the plug and circlip. Pull the fork spring out of the stanchion. On some models two springs are used, separated by a spacer.

5   Remove the fork leg from the vice and prise the dust cover off the lower leg, sliding it up off the stanchion. The stanchion is now free to be pulled from the lower leg. Invert the stanchion and allow the damper rod assembly to slide out. The lower end of the damper rod is fitted with a seating piece, which will remain either on the rod or fall off into the fork lower leg when the stanchion is withdrawn.

6   The fork lower leg is fitted with an oil seal retained by a circlip. Remove the circlip and oil seal **ONLY** if the seal is damaged, as it will almost certainly be damaged when being prised from position.

4   Front forks: examination and renovation

1   The parts most liable to wear over an extended period of service are the wearing surfaces of the fork stanchion and lower leg, the damper assembly within the fork tube and the oil seal at the sliding joint. Wear is normally accompanied by a tendency for the forks to judder when the front brake is applied and it should be possible to detect the increased amount of play by pulling and pushing on the handlebars when the front brake is applied fully. This type of wear should not be confused with slack steering head bearings, which can give identical results.

2   Renewal of the worn parts is quite straightforward. Particular care is necessary when renewing the oil seal, to ensure that the feather edge seal is not damaged during reassembly. Both the seal and the fork tube should be greased, to lessen the risk of damage.

3   After an extended period of service, the fork springs may take a permanent set. If there is any doubt as to their condition check the free lengths against those of a new spring. If there is a noticeable difference, renew the springs as a complete set.

4   Check the outer surface of the fork tube for scratches or roughness. It is only too easy to damage the oil seal during reassembly, if these high spots are not eased down. The fork tubes are unlikely to bend unless the machine is damaged in an accident. Any significant bend will be detected by eye, but if there is any doubt about straightness, roll the tubes on a flat surface. If the tubes are bent, they must be renewed. Unless specialised repair equipment is available, it is rarely practicable to straighten them to the necessary standard.

5   The dust seals must be in good order if they are to fulfil their proper function. Replace any that are split or damaged.

6   Damping is effected by the damper units contained within each fork tube. The damping action can be controlled within certain limits by changing the viscosity of the oil used as the damping medium, although a change is unlikely to prove necessary except in extremes of climate.

7   Note that the forks are not fitted with renewable bushes. If wear develops, the stanchions and/or the lower fork legs will have to be renewed.

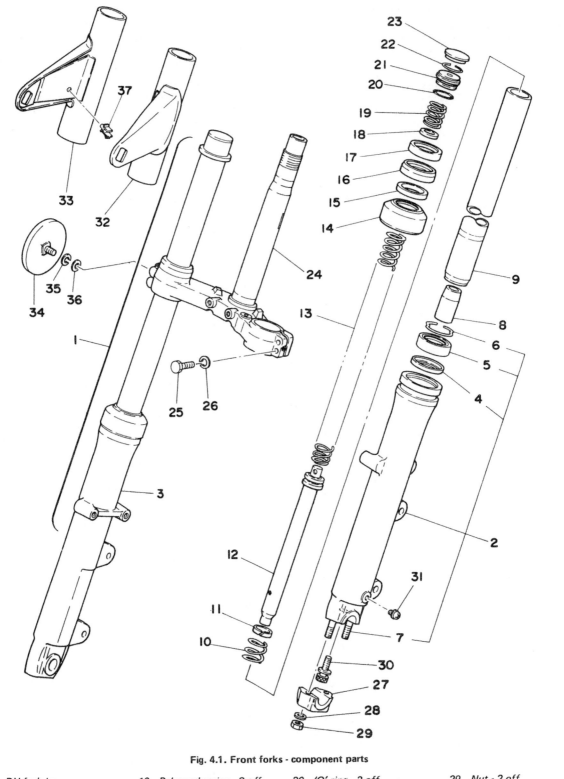

Fig. 4.1. Front forks - component parts

1	RH fork leg	10	Rebound spring - 2 off	20	'O' ring - 2 off	29	Nut - 2 off
2	LH lower leg	11	Damper ring - 2 off	21	Plug - 2 off	30	Socket bolt - 2 off
3	RH lower leg	12	Damper rod - 2 off	22	Circlip - 2 off	31	Drain plug - 2 off
4	Seal seat - 2 off	13	Spring - 2 off	23	Cover - 2 off	32	LH shroud
5	Oil seal - 2 off	14	Dust excluder - 2 off	24	Lower yoke/steering	33	RH shroud
6	Circlip - 2 off	15	Seal - 2 off		stem	34	Reflector - 2 off
7	Stud - 2 off	16	Shroud lower seat - 2 off	25	Bolt - 4 off	35	Spring washer - 2 off
8	Damper rod seat - 2 off	17	Shroud upper seat - 2 off	26	Spring washer - 4 off	36	Plain washer - 2 off
9	Stanchion (upper tube) -	18	Spring upper seat - 2 off	27	Spindle clamp	37	Clip - 2 off
	2 off	19	Spring - 2 off	28	Washer - 2 off		

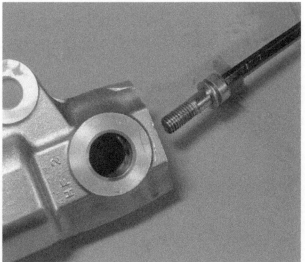

3.3 Remove the damper rod retaining screw

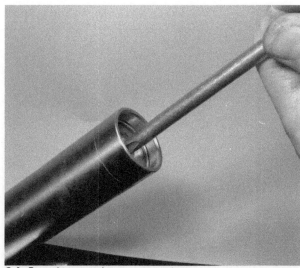

3.4a Press down on plug to remove circlip

3.4b Withdraw the plug and ...

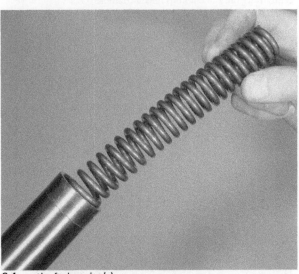

3.4c ... the fork spring(s)

3.5a Prise the dust cap off the lower leg and ...

3.5b ... separate the stanchion from the leg

3.5c Invert the stanchion to remove damper rod

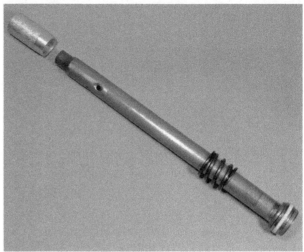

3.5d Damper rod assembly, including lower seat

3.6 Oil seal is retained by a circlip

5 Steering head bearings: examination and renovation

1   Before reassembling and refitting the front forks, the steering head bearings should be checked for wear. Clean the inner and outer races thoroughly in petrol before inspection. Check the rollers and the bearing tracks for indentation or pitting or in extreme cases for fracture of the outer races.

2   The bearing outer races are a tight drive fit in the head lug casting. If their condition dictates renewal, they may be driven from position using a long handled drift. The lower bearing inner race is a push fit on the steering stem and may be levered from place.

3   Before reassembling the fork yoke sub-assembly, grease the steering head bearings thoroughly with a multi-purpose grease. There is no provision for further lubrication after assembly.

6 Front forks: replacement

1   Replace the front forks by following in reverse the dismantling procedures described in Sections 2 and 3 of this Chapter. When replacing the fork springs note that they are of variable

pitch. The more widely spaced coils should be towards the **bottom** of the fork leg.

2   Before replacing the spring retaining plugs, check that the drain plugs are in position, and replenish each leg with 170 cc (5.75/4.8 US/Imp fl oz) of good quality SAE 20 fork oil.

3   If the fork stanchions prove difficult to re-locate through the fork yokes, make sure their outer surfaces are clean and polished so that they will slide more easily. It is often advantageous to apply a small quantity of soap liquid to the stanchion, to help insertion through the rubber shroud seats.

4   The annular groove near the top of each fork stanchion should be aligned with the top of the upper yoke, to ensure that the correct rake and trail is maintained. Before fully tightening the front wheel spindle clamps and the fork yoke pinch bolts, bounce the forks several times to ensure they work freely and are clamped in their original settings. Complete the final tightening from the wheel spindle clamps upward.

4   Before the machine is used on the road, check the adjustment of the steering head bearings. If they are too slack, judder will occur. There should be no detectable play in the head races when the handlebars are pulled and pushed, with the front brake applied hard.

5   Overtight head races are equally undesirable. It is possible to unwittingly apply a loading of several tons on the head bearings by overtightening, even though the handlebars appear to turn quite freely. Overtight bearings will cause the machine to roll at low speeds and give generally imprecise handling with a tendency to weave. Adjustment is correct if there is no perceptible play in the bearings and the handlebars will swing to full lock in either direction, when the machine is on the centre stand with the front wheel clear of the ground. Only a slight tap should cause the handlebars to swing. After adjustment has been made by use of the lower peg nut on the steering stem, the upper nut should be tightened, followed by the pinch bolt to the rear of the upper yoke.

7 Steering head lock

1   The steering head lock is attached to the left-hand side of the steering head. It is retained by a rivet. When in a locked position, the plunger extends and engages with a portion of the steering head stem, so that the handlebars are locked in position and cannot be turned.

2   If the lock malfunctions, it must be renewed. A repair is impracticable. When the lock is changed it follows that the key must be changed too, to correspond with the new lock.

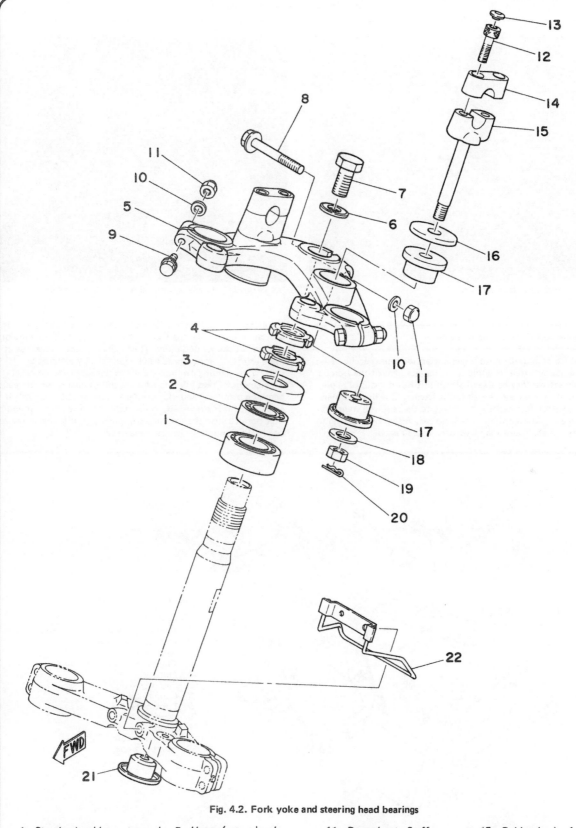

Fig. 4.2. Fork yoke and steering head bearings

1	Steering head lower tapered roller bearing	5	Upper (crown) yoke
2	Steering head upper tapered roller bearing	6	Plain washer
3	Cover	7	Crown bolt
4	Peg nut - 2 off	8	Pinch bolt

1 Steering head lower tapered roller bearing
2 Steering head upper tapered roller bearing
3 Cover
4 Peg nut - 2 off
5 Upper (crown) yoke
6 Plain washer
7 Crown bolt
8 Pinch bolt
9 Pinch bolt - 2 off
10 Plain washer - 3 off
11 Domed nut - 3 off
12 Socket bolt - 4 off
13 Cap - 4 off
14 Handlebar clamp - 2 off
15 Handlebar holder - 2 off
16 Plain washer - 2 off
17 Rubber bush - 4 off
18 Plain washer - 2 off
19 Nut - 2 off
20 Spring clip - 2 off
21 Plug
22 Cable guide

6.2 Replenish each fork leg with oil

6.4 Groove in stanchion must align with top of yoke

8 Frame: examination and renovation

1   The frame is unlikely to require attention unless accident damage has occurred. In some cases, replacement of the frame is the only satisfactory course of action if it is badly out of alignment. Only a few frame repair specialists have the jigs and mandrels necessary for resetting the frame to the required standard of accuracy and even then there is no easy means of assessing to what extent the frame may have been overstressed.

2   After the machine has covered a considerable mileage, it is advisable to examine the frame closely for signs of cracking or splitting at the welded joints. Rust can also cause weakness at these joints. Minor damage can be repaired by welding or brazing, depending on the extent and nature of the damage.
3   Remember that a frame which is out of alignment will cause handling problems and may even promote 'speed wobbles'. If misalignment is suspected, as the result of an accident, it will be necessary to strip the machine completely so that the frame can be checked and, if necessary, renewed.

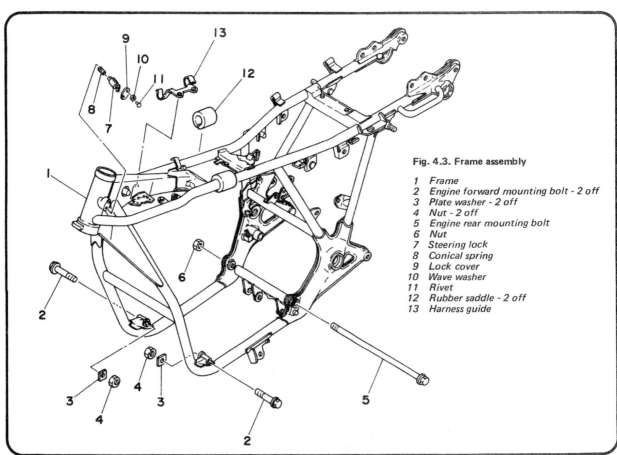

Fig. 4.3. Frame assembly

1   Frame
2   Engine forward mounting bolt - 2 off
3   Plate washer - 2 off
4   Nut - 2 off
5   Engine rear mounting bolt
6   Nut
7   Steering lock
8   Conical spring
9   Lock cover
10 Wave washer
11 Rivet
12 Rubber saddle - 2 off
13 Harness guide

9 Swinging arm bearing: checking and adjustment

1   The rear swinging arm fork pivots on two tapered roller bearings, which are supported on adjustable screw stubs fitted to the lugs either side of the frame. After a period of time the tapered roller bearings will wear slightly, allowing a small amount of lateral shake at the rear wheel. This condition, even in its early stages, will have a noticeable effect on handling.
2   To check the play accurately, and if necessary to make suitable adjustments, it will be necessary to remove the rear wheel and detach the rear suspension units.
3   Place the machine on the centre stand so that the rear wheel is clear of the ground. In order to remove the rear wheel spindle without first detaching the right-hand silencer, the rear suspension units must be compressed slightly by applying weight to the rear of the machine. A short length of steel cable - supplied in the tool kit - may then be attached, on hooks provided, between the right-hand swinging arm member and the upper suspension unit mounting lug.
4   Remove the rear wheel spindle nut, which is secured by a split pin. Slacken the clamp bolt from the end of the right-hand swinging arm member and withdraw the wheel spindle. Rotate the caliper unit upwards on its upper pivot bolt so that the pads clear the disc, and lift the caliper support bracket up so that it rests on the lug provided on the end of the swinging arm right-hand member. Note the spacer which is placed between the caliper and support bracket. Lift the wheel across to the right and off the final drive box splines. The rear mudguard is hinged at a point to the rear of the dualseat, which allows easy removal of the rear wheel. Unscrew the two hinge bolts, raise the hinged piece and push the bolts in again to support the raised mudguard flap.
5   Detach the rear suspension units at their lower ends by removing the pivot bolts.
6   The swinging arm fork can now be checked for play. Grasp the fork at the rear end and push and pull firmly in a lateral direction. Any play will be magnified by the leverage effect. Move the swinging arm up and down as far as possible. Any roughness or a tightness at one point may indicate bearing damage. If this is suspected, the bearings should be inspected after removal of the swinging arm as described in the following Section.
7   If play is evident, remove the plastic cover plugs from the adjuster stubs either side of the frame and slacken off the adjuster locknuts. Using a vernier gauge, check that the distances between each end of the swinging arm cross-member and the adjacent part of the frame are within 1.6 mm (0.062 in) of each other. If the difference is greater than this, slacken the adjuster stub ½ of a turn on the side with the larger gap, and then tighten the opposite stub to 0.5 - 0.6 kg m (43 - 52 in lbs). Check the

gap again and if necessary, readjust, using the same procedure. When using a torque wrench the load on the bearings is adjusted automatically so that all play is taken up. Where no torque wrench is available, centralise the swinging arm and then tighten the stubs an equal amount, about 1/8 of a turn at a time. Whilst tightening, move the swinging arm up and down until it can be felt that the bearings are beginning to drag. It is at this point that adjustment is correct.
8   After adjustment is completed, tighten the locknuts without allowing the stubs to rotate. Refit the plastic plugs.
9   Reassemble the rear wheel and suspension units by reversing the dismantling procedure.

10 Swinging arm: removal, renovation and replacement

1   If on inspection for play in the swinging arm bearings it is found that damage has occurred, or wear is excessive, the swinging arm fork should be removed. Commence by following paragraphs 3 - 5 of the preceding Section and then detach the swinging arm as follows:
2   Prise the final driveshaft rubber gaiter off the boss to the rear of the gearbox so that access to the final driveshaft flange is gained. Slacken evenly and remove the four flange bolts, turning the flange as necessary to reach the bolts.
3   The final drive box may be detached as a complete unit after removing the four flange nuts and washers. Support the weight of the casing as the nuts are removed. Drainage of the lubricating oil is not required, provided the casing is moved and stored in an upright position.
4   Remove the pivot nut washers which secure the rear of the caliper support bracket. The nut is secured by a split pin. The complete caliper/support bracket may be lifted towards the front of the machine and tied to a suitable portion of the frame.
5   Loosen the locknuts on the swinging arm adjuster stubs after detaching the plastic covers from either side of the frame. Unscrew the stubs completely and lift the swinging arm fork, complete with final driveshaft, from the machine.
6   To each end of the swinging arm cross-member is fitted a spacer, oil seal and tapered roller bearing. The spacer is a push fit in the oil seal. To inspect or remove the bearing on either side, the seal must be prised from position. Removal will almost certainly damage the seal and a new one will therefore be required. Lever out the seals with a screwdriver. Take out the bearing inner races and clean and inspect them. Clean the bearing outer races whilst they are still in place. Check the rollers for pitting and the outer race for pitting or indentation. If the bearings need renewing, the inner races may be levered from position.
7   When assembling, clean and lubricate the bearings. Use a waterproof grease of the type recommended for wheel bearings.
8   Refit the swinging arm by reversing the dismantling procedure. Adjust the bearings by referring to Section 9, paragraphs 7 and 8.

9.4a Remove split pin and nut

9.4b Slacken clamp bolt and withdraw the spindle

9.4c Remove caliper bracket spacer and ...

9.4d ... swing caliper up, off the disc

9.4e Remove the wheel towards the rear

9.5 Detach the suspension units at lower ends

9.7 Use Allen key to adjust swinging arm play

10.3 Final drive box is held by four nuts

10.4a Caliper pivots on studs, secured by nut and split pin

10.4b Remove the adjuster pivot studs and ...

10.4c ... lift swinging arm fork out rearwards

10.6a Displace the spacer and ...

10.6b ... prise out the oil seal to enable ...

10.6c ... removal of bearing inner race

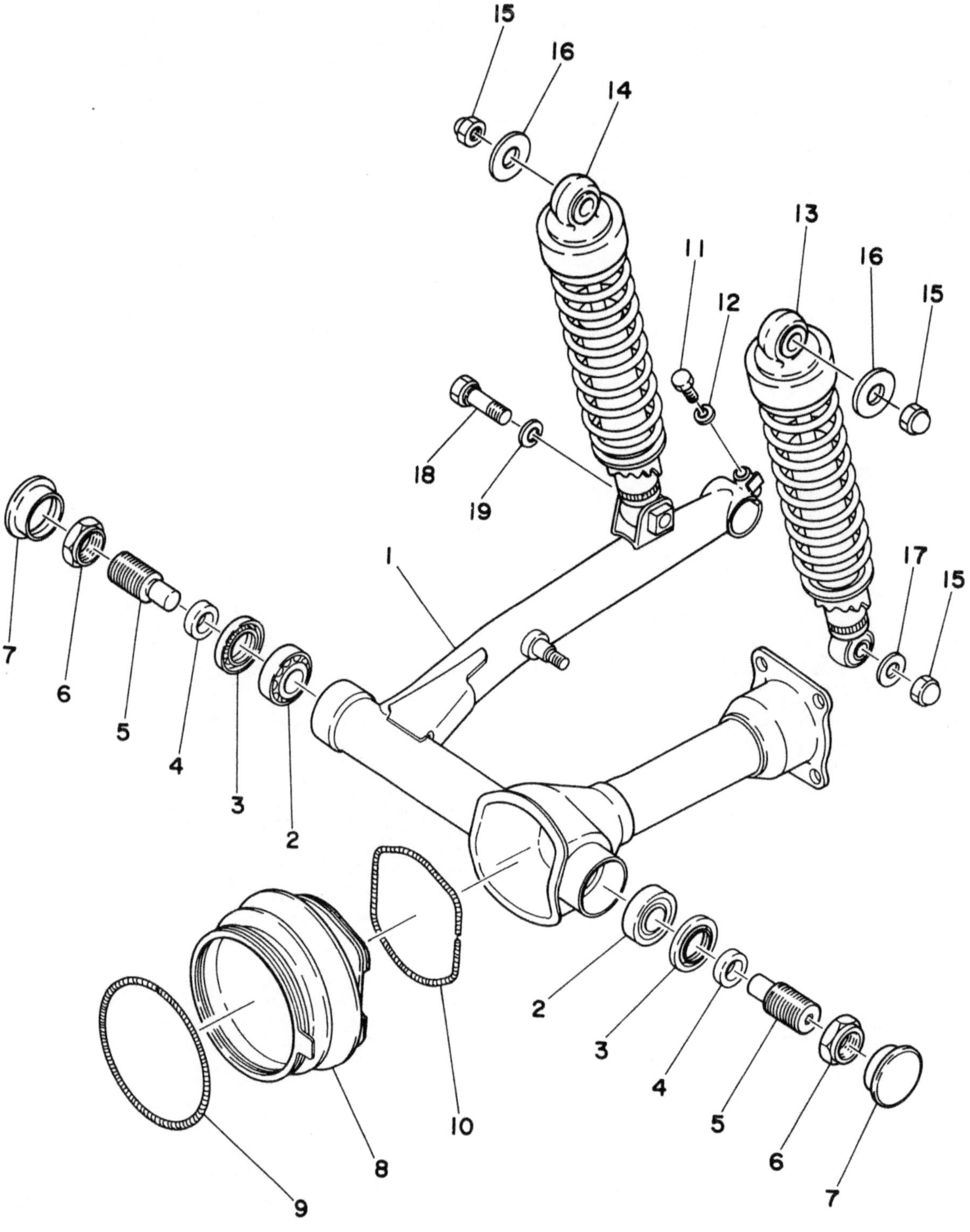

Fig. 4.4. Rear swinging arm assembly

1	Swinging arm fork	5	Adjuster pivot stub - 2 off
2	Tapered roller bearing - 2 off	6	Locknut - 2 off
3	Oil seal - 2 off	7	Cap - 2 off
4	Collar - 2 off	8	Gaiter
9	Spring gaiter	14	RH rear suspension unit
10	Spring gaiter	15	Domed nut - 3 off
11	Bolt	16	Plain washer - 2 off
12	Spring washer	17	Plain washer
13	LH rear suspension unit	18	Bolt
		19	Spring washer

11 Final driveshaft: examination and renewal

1 In due course the joint on the upper end of the driveshaft will wear. The joint is of the constant velocity type, sealed for life with its own lubrication reservoir and hence should have a long life expectancy. The final driveshaft and joint may be removed from the swinging arm unit, after the latter component has been removed from the machine.

2 Using two screwdrivers as levers with the lever ends placed behind the splined lower boss, withdraw the shaft from the swinging arm. The CV joint may be lifted from the forward end. Wear of the joint will be self-evident, giving a rough notchy feel when it is flexed.

3 Check the condition of the splines at both ends of the shaft. If damage is evident, the shaft should be renewed. Inspect the condition of the seal and renew it, if required.

4 When refitting the shaft, lubricate the splined ends with graphite grease.

11.3a Check condition of splined boss on shaft and ...

11.3b ... mating splines in drive box input shaft

11.3c Grease all splines before reassembly

12 Final drive bevel gear: examination and renovation

1 Dismantling the bevel drive is beyond the scope of this manual, and the majority of amateur mechanics. Wear or damage may be indicated by a high pitched whine. Backlash between the crownwheel and pinion may be assessed by holding the output splined boss and rotating the input shaft, first one way and then the other.

2 Failure of the seals either at the input shaft or output shaft will be self-evident by oil leakage.

3 Check the splines on the input shaft and output boss for wear or damage.

4 If attention to the final drive bevel box is required, the complete unit should be returned to a Yamaha Service agent, who will have the necessary tools and experience to carry out inspection and overhaul.

13 Rear suspension units: examination

1 The rear suspension units fitted to the Yamaha XS 750

model are of the normal hydraulically damped type, adjustable to give 5 different spring settings. A C spanner included in the tool kit should be used to turn the lower spring seat and so alter its position on the adjustment projection. When the spring seat is turned so that the effective length of the spring is shortened the suspension will become heavier.

2 If a suspension unit leaks, or if the damping efficiency is reduced in any other way the two units must be replaced as a pair. For precise roadholding it is imperative that both units react to movement in the same way. It follows that the units must always be set at the same spring loading.

14 Centre stand: examination

1 The centre stand is attached to the machine by two bolts on the bottom of the frame. It is returned by a centre spring. The bolts and spring should be checked for tightness and tension respectively. A weak spring can cause the centre stand to 'ground' on corners and unseat the rider.

15 Prop stand: examination

1   The prop stand is secured to a plate on the frame with a bolt and nut, and is retracted by a tension spring. Make sure the bolt is tight and the spring not overstretched, otherwise an accident can occur if the stand drops during cornering.

16 Footrests: examination and renovation

1   Each footrest is an individual unit retained by a single bolt to a suitable part of the frame.
2   Both pairs of footrests are pivoted on clevis pins and spring loaded in the down position. If an accident occurs, it is probable that the footrest peg will move against the spring loading and remain undamaged. A bent peg may be detached from the mounting, after removing the clevis pin securing split pin and the clevis pin itself. The damaged peg can be straightened in a vice, using a blowlamp flame to apply heat at the area where the bend occurs. The footrest rubber will, of course, have to be removed as the heat will render it unfit for service.

17 Rear brake pedal: examination and renovation

1   The rear brake pedal pivots on a shaft which passes through the frame right-hand intersection lug. The shaft carrying the brake arm is splined, to engage with splines of the rear brake pedal. The pedal is retained to the shaft by a simple pinch bolt arrangement.
2   If the brake pedal is bent or twisted in an accident, it should be removed by slackening the pinch bolt and straightened in a manner similar to that recommended for the footrests in the preceding Section.
3   Make sure the pinch bolt is tight. If the lever is a slack fit on the splines, they will wear rapidly and it will be difficult to keep the lever in position.

18 Dualseat: removal and replacement

1   The dualseat is attached to the right-hand frame tube by means of two pivots on which it hinges. A catch on the left-hand frame tube locks the dualseat in position, under normal riding conditions.
2   To release the dualseat from the machine, lift the catch and lift the dualseat so that the pivots on the right-hand side are exposed. Disconnect the support rod from the slide way mounted above the mudguard.
    If the split pins though the pivots are removed and the pivot pins withdrawn, the dualseat can be lifted away.

19 Speedometer and tachometer heads: removal and replacement

1   The tachometer and speedometer heads are rubber mounted on a common support bracket, which is retained on the fork top (crown) yoke by two integral studs secured on the underside by nuts.
2   The instrument heads may be detached from the mounting plate individually, after the cable has been detached by unscrewing the knurled rings and the two dome nuts and washers removed from the underside.
3   Each instrument head rests in a shell where it is supported by a rubber cushioning ring and secured by two dome nuts. Remove the nuts and washers and separate the instrument from the shell. It will be necessary to remove the bulbs from the case of each instrument head by pulling the bulbholders from their seatings; each is retained by a rubber cup.
4   Do not misplace the rubber cushion interposed between the mounting bracket and the instrument case to damp out the undesirable effects of vibration.
5   Apart from defects in either the drive or the drive cable, a speedometer or tachometer that malfunctions is difficult to repair. Fit a new one, or alternatively entrust the repair to a component instrument repair specialist.
6   Remember that a speedometer in correct working order is a statutory requirement in the UK and many other countries. Apart from this legal requirement, reference to the odometer reading is the best means of keeping in pace with the maintenance schedule.

20 Speedometer and tachometer drive cables: examination and maintenance

1   It is advisable to detach both cables from time to time in order to check whether they are lubricated adequately, and whether the outer coverings are compressed or damaged at any point along their run. Jerky or sluggish movements can often be attributed to a cable fault.
2   For greasing, withdraw the inner cable. After wiping off the old grease, clean with a petrol-soaked rag and examine the cable for broken strands or other damage.
3   Regrease the cable with high melting point grease, taking care not to grease the last six inches at the point where the cable enters the instrument head. If this precaution is not observed, grease will work into the head and immobilise the movement.
4   If either instrument ceases to function, suspect a broken cable. Inspection will show whether the inner cable has broken; if so, the inner cable alone can be renewed and reinserted in the outer casing, after greasing. Never fit a new inner cable alone if the outer covering is damaged or compressed at any point.

21 Speedometer and tachometer drive: location and examination

1   The speedometer drive gearbox is fitted on the front wheel spindle where it is driven internally by the left-hand side of the wheel hub.
2   The gearbox rarely gives trouble if it is lubricated with grease at regular intervals. This can only be done after the wheel has been removed and the gearbox has been detached since no external grease nipple is fitted. The gearbox can be pulled from position after wheel removal.
3   The tachometer is driven from a scroll gear on the exhaust cam, via a driveshaft which is a push fit in the front of the cylinder head. The driveshaft and housing are retained by a forked plate, held by a single screw. Failure of the drive gear is unlikely, due to the good operating conditions.

22 Sidecar alignment

1   Using specialized fittings, little difficulty is experienced when attaching a sidecar to a Yamaha XS 750 model. It should be remembered, however, that Yamaha do not recommend the fitting of a sidecar, and so the warranty of the machine may be invalidated by so doing.
2   Good handling characteristics of the outfit will depend on the accuracy with which the sidecar is aligned. Provided the toe-in and lean-out are within prescribed limits, good handling characteristics should result, leaving scope for other minor adjustments about which opinions vary quite widely.
3   To check the toe-in, check that the front and rear wheels of the motorcycle are correctly in line and adjust the sidecar fittings so that the sidecar wheel is approximately parallel to a line drawn between the front and rear wheels of the machine. Re-adjust the fittings so that the sidecar wheel has a slight toe-in toward the front wheel of the motorcycle, as shown in Fig. 4.5A. When the amount of toe-in is correct, the distance 'B'

should be from 3/8 to ¾ in less than the distance at 'A'.

4   Lean-out is checked by attaching a plumb line to the handle-bars and measuring the distance between 'C' and 'D' as shown in Fig. 4.5B. Lean-out is correct when the distance 'C' is approximately 1 inch greater than at 'D'.

5   Aligning a sidecar correctly is a task for the specialist. It is always advisable to seek the advice of a sidecar specialist as these notes are provided in very general terms, to act as a guide only.

23 Cleaning the machine

1   After removing all surface dirt with warm water and a rag or sponge, use a cleaning compound such as 'Gunk' or 'Jizer' for the oily parts. Apply the cleaner with a brush when the parts are dry so that it has an opportunity to soak into the film of oil or grease. Finish off by washing down liberally, taking care that water does not enter into the carburettors, air cleaners or electrics. If desired, a polish such as Solvol Autosol can be applied to the alloy parts to give them full lustre. Application of a wax polish to the cycle parts and a good chrome cleaner to the chrome parts will also give a good finish. Always wipe the machine down if used in the wet. There is less chance of water getting into control cables if they are regularly lubricated, which will prevent stiffness of action.
action.

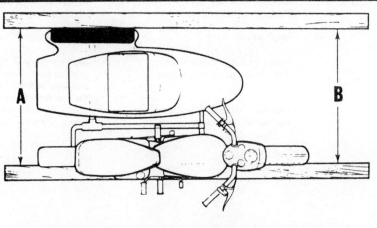

Fig. 4.5a. Aligning the sidecar wheel to the correct amount of toe-in

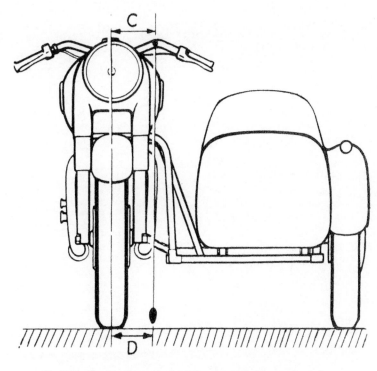

Fig. 4.5b. Setting the amount of 'lean out' by using a plumb line

24 Fault diagnosis: Frame and forks

Symptom	Cause	Remedy
Machine veers to the left or right with hands off handlebars	Wheels out of alignment Forks twisted Frame bent	Check and realign. Strip and repair. Strip and repair or renew.
Machine tends to roll at low speeds	Steering head bearing not adjusted correctly or worn	Check adjustment and renew bearings if necessary.
Machine tends to wander	Worn swinging arm bearings	Check and renew bearings. Check adjustment and renew if necessary.
Forks judder when front brake applied	Steering head bearings slack Fork components worn	Adjust bearings. Strip forks, and renew all worn parts.
Forks bottom	Short of oil	Replenish with correct viscosity oil.
Fork action stiff	Fork legs out of alignment Bent shafts, or twisted ie. yokes	Slacken clamp bolts, front wheel spindle and top bolts. Pump forks several times and tighten from bottom upwards. Strip and renew parts, if damaged.
Machine pitches badly	Defective rear suspension units or ineffective fork damping	Check damping action. Check grade and quantity of oil in front forks.

Chapter 5 Wheels, brakes and tyres

Contents

Specifications

Tyres
Size:

Front ...	...	...	...	...	...	...	...	...	19 x 3.25 in H19
Rear ...	...	...	...	...	...	...	...	...	18 x 4.00 in H18

Tyre pressures

									Solo	*Pillion
Front ...	...	...	...	...	...	...	...	...	26 psi (1.8 kg cm^2)	28 psi (2.0 kg cm^2)
Rear ...	...	...	...	...	...	...	...	...	28 psi (2.0 kg cm^2)	33 psi (2.3 kg cm^2)

*Also for continuous riding at speeds in excess of 60 mph.

Brakes

Front ...	...	...	...	...	...	...	...	...	Hydraulic 270 mm twin discs
Rear ...	...	...	...	...	...	...	...	...	Hydraulic 270 mm single disc

Hydraulic fluid

Type ...	...	...	...	...	...	...	...	...	DOT 3 (USA), SAE J1703 (UK)

1 General description

All but the very early Yamaha XS 750 models are fitted with cast aluminium 7-spoke wheels at both the front and rear, of which the former is of 19 inch diameter and the latter 18 inch. The front tyre section is 3.25 in and that of the rear 4.00 inch. The original equipment tyres are manufactured by Bridgestone and have an H speed rating, safe up to 130 mph. Twin disc brakes are fitted at the front and a single disc at the rear. All discs are of 270 mm diameter. The front and rear calipers, although not interchangeable, are of fundamentally the same design and type.

2 Front wheel: examination and renovation

1 Carefully check the complete wheel for cracks and chipping, particularly at the spoke roots and the edge of the rim. As a general rule a damaged wheel must be renewed as cracks will cause stress points which may lead to sudden failure under heavy load. Small nicks may be radiused carefully with a fine

file and emery paper (No. 600 - No. 1000) to relieve the stress. If there is any doubt as to the condition of a wheel, advice should be sought from a Yamaha repair specialist.
2 Each wheel is covered with a coating of lacquer, to prevent corrosion. If damage occurs to the wheel and the lacquer finish is penetrated, the bared aluminium alloy will soon start to corrode. A whitish grey oxide will form over the damaged area, which in itself is a protective coating. This deposit however, should be removed carefully as soon as possible and a new protective coating of lacquer applied.
3 Check the lateral run out at the rim by spinning the wheel and placing a fixed pointer close to the rim edge. If the maximum run out is greater than 2.0 mm (0.080 in), Yamaha recommend that the wheel be renewed. This is, however, a counsel of perfection; a run out somewhat greater than this can probably be accommodated without noticeable effect on steering. No means is available for straightening a warped wheel without resorting to the expense of having the wheel skimmed on all faces. If warpage was caused by impact during an accident, the safest measure is to renew the wheel complete. Worn wheel bearings may cause rim run out. These should be renewed as described in Section 9 of this Chapter.

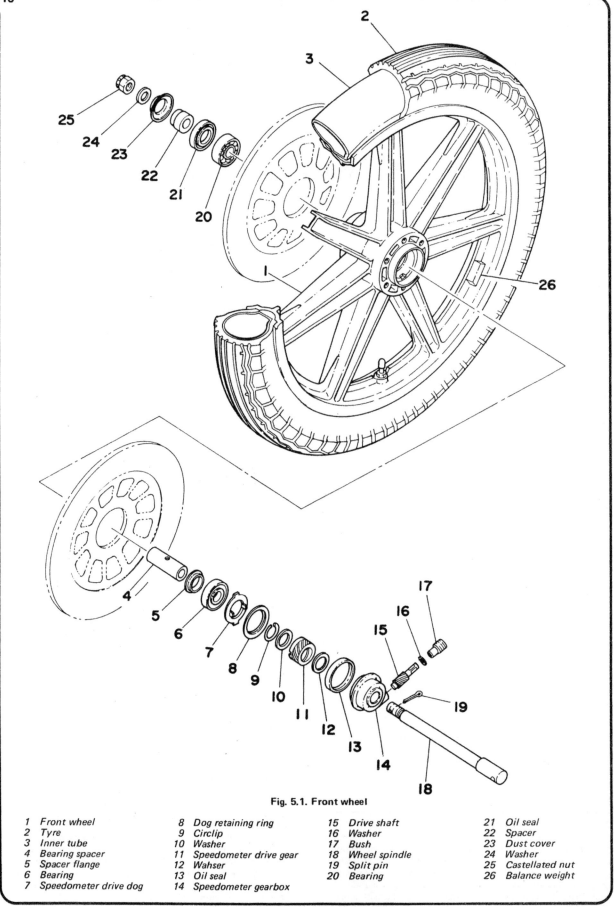

Fig. 5.1. Front wheel

1	Front wheel	8	Dog retaining ring	15	Drive shaft	21	Oil seal
2	Tyre	9	Circlip	16	Washer	22	Spacer
3	Inner tube	10	Washer	17	Bush	23	Dust cover
4	Bearing spacer	11	Speedometer drive gear	18	Wheel spindle	24	Washer
5	Spacer flange	12	Wahser	19	Split pin	25	Castellated nut
6	Bearing	13	Oil seal	20	Bearing	26	Balance weight
7	Speedometer drive dog	14	Speedometer gearbox				

3 Front disc brake: checking and renewing the pads

1 To facilitate the checking of brake pad wear, each caliper is provided with an inspection window closed by a small cover. Prise the cover from position and inspect both pads. Each pad has a red wear limit line around its periphery. If either pad has worn down to or past the line, both pads in that set should be renewed. In practice, it is probable that both sets of pads will wear at a similar rate, and therefore the two sets will require renewal at the same time.

2 Removal of the pads is straightforward. The procedure is identical for both calipers, and does not require the hydraulic hose to be disconnected. Remove the two bolts that secure the caliper bracket to front fork leg, and lift the complete unit up off the brake disc. Remove the single bolt which secures the caliper unit to the caliper mounting bracket. Unscrew the crosshead screw from the inner face of the caliper, noting that the screw acts as a detent for the pads. Pull the support bracket from the main caliper and lift both pads from position. Note the various shims and their positions, before removal.

3 Fit new pads by reversing the dismantling sequence. If difficulty is encountered when fitting the caliper over the brake disc, due to the reduced distance between the new pads, use a wooden lever to push the pad on the piston side inwards.

4 In the interests of safety, always check the function of the brakes before taking the machine on the road.

4 Front disc brake: removing, renovating and replacing the caliper units

1 Before either caliper assembly can be removed from the fork leg upon which it is mounted, it is first necessary to drain off the hydraulic fluid. Disconnect the brake pipe at the union connection it makes with the caliper unit and allow the fluid to drain into a clean container. It is preferable to keep the front brake lever applied throughout this operation, to prevent the fluid from leaking out of the reservoir. A thick rubber band cut from a section of inner tube will suffice, if it is wrapped tightly around the lever and the handlebars.

2 Note that brake fluid is an extremely efficient paint stripper. Take care to keep it away from any paintwork on the machine or from any clear plastic, such as that sometimes used for instrument glasses.

3 When the fluid has drained off, remove the caliper mounting bolts, separate the two main caliper components and remove the pads as described in the preceding Section.

4 To displace the piston, apply a blast of compressed air to the brake fluid inlet. Take care to catch the piston as it emerges from the bore - if dropped or prised out with a screwdriver

a piston may suffer irreparable damage. Before removing the piston, displace the dust seal which is retained by a circlip.

5 Remove the sleeve and protective boot upon which the caliper unit slides. If play has developed between the sleeve and the caliper, the former must be renewed. Check the condition of the boot, renewing it if necessary.

6 The parts removed should be cleaned thoroughly, using only brake fluid as the liquid. Petrol, oil or paraffin will cause the various seals to swell and degrade, and should not be used under any circumstances. When the various parts have been cleaned, they should be stored in polythene bags until reassembly, so that they are kept dust free.

7 Examine the pistons for score marks or other imperfections. If they have any imperfections they must be renewed, otherwise air or hydraulic fluid leakage will occur, which will impair braking efficiency. With regard to the various seals, it is advisable to renew them all, irrespective of their appearance. It is a small price to pay against the risk of a sudden and complete front brake failure. It is standard Yamaha practice to renew the seals every two years, even if no braking problems have occurred.

8 Reassemble under clinically-clean conditions, by reversing the dismantling procedure.

Apply a small quantity of graphite grease to the slider sleeve before fitting the boot. Reconnect the hydraulic fluid pipe and make sure the union has been tightened fully. Before the brake can be used, the whole system must be bled of air, by following the procedure described in Section 7 of this Chapter.

3.2a Caliper bracket is retained by two bolts

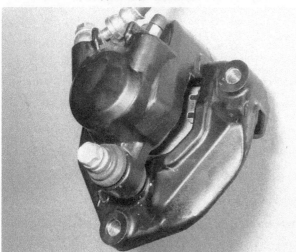

3.2b Remove central bolts to separate caliper from support

4.4 Piston dust seal is retained by a circlip

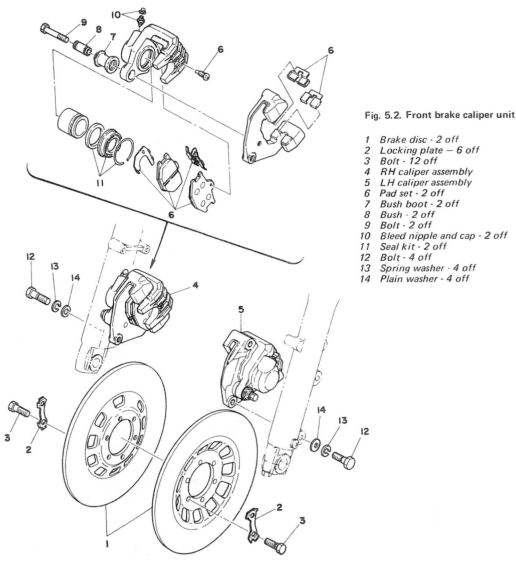

Fig. 5.2. Front brake caliper unit

1   Brake disc - 2 off
2   Locking plate — 6 off
3   Bolt - 12 off
4   RH caliper assembly
5   LH caliper assembly
6   Pad set - 2 off
7   Bush boot - 2 off
8   Bush - 2 off
9   Bolt - 2 off
10   Bleed nipple and cap - 2 off
11   Seal kit - 2 off
12   Bolt - 4 off
13   Spring washer - 4 off
14   Plain washer - 4 off

5   Master cylinder: examination and renewing seals

1   The master cylinder and hydraulic fluid reservoir takes the form of a combined unit mounted on the right-hand side of the handlebars, to which the front brake lever is attached.

2   Before the master cylinder unit can be removed and dismantled, the system must be drained. Place a clean container below each brake caliper unit and attach a plastic tube from the bleed screw of each caliper unit to the container. Lift off the master cylinder cover (cap), gasket and diaphragm, after removing the four countersunk retaining screws. Open the bleed screws one complete turn and drain the system by operating the brake lever until the master cylinder reservoir is empty. Close the bleed screws and remove the tube.

3   Before dismantling the master cylinder, it is essential that a clean working area is available on which the various component parts can be laid out. Use a sheet of white paper, so that none of the smaller parts can be overlooked.

4   Disconnect the stop lamp switch and front brake lever, taking care not to misplace the brake lever return spring. The stop lamp switch is a push fit in the lever stock. The lever pivots on a bolt retained by a single nut. Remove the brake hose by unscrewing the banjo union bolt. Take the master cylinder away from the handlebars by removing the two bolts that clamp it to the handlebars. Take care not to spill

any hydraulic fluid on the paintwork or on plastic or rubber components.

5   Withdraw the rubber boot that protects the end of the master cylinder and remove the snap ring that holds the piston assembly in position, using a pair of circlip pliers. The piston assembly can now be drawn out, followed by the return valve, spring cup and return spring.

6   The spring cup can now be separated from the end of the return valve spring and the main cup prised off the piston.

7   Examine the piston and the cylinder cup very carefully. If either is scratched or has the working surface impaired in any other way, it must be renewed without question. Reject the various seals, irrespective of their condition, and fit new ones in their place. It often helps to soften them a little before they are fitted by immersing them in a container of clean brake fluid.

8   When reassembling, follow the dismantling procedure in reverse, but take great care that none of the component parts is scratched or damaged in any way. Use brake fluid as the lubricant whilst reassembling. When assembly is complete, reconnect the brake fluid pipe and tighten the banjo union bolt.

9   Use two new sealing washers at the union so that the banjo bolt does not require overtightening to effect a good seal. Refill the master cylinder with DOT 3 or SAE J1703 brake fluid and bleed the system of air by following the procedure described in Section 7 of this Chapter.

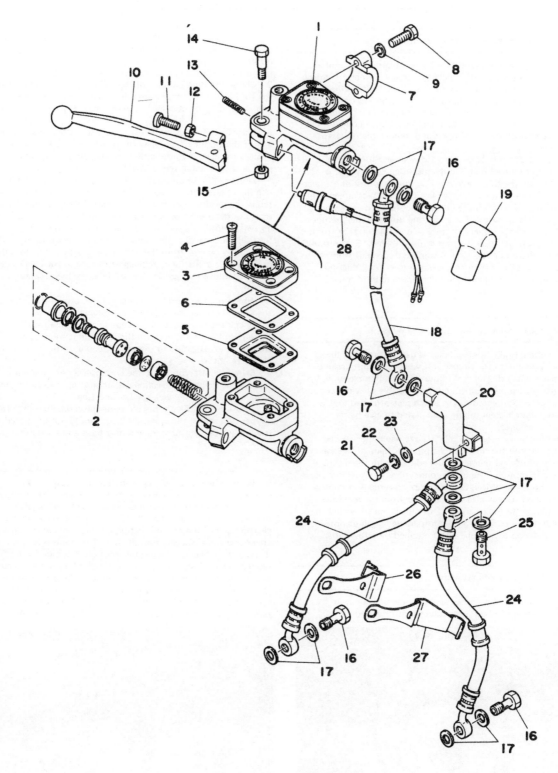

Fig. 5.3. Front brake master cylinder

1 Master cylinder assembly	8 Bolt - 2 off	15 Nut	22 Spring washer
2 Piston kit	9 Spring washer - 2 off	16 Banjo bolt - 4 off	23 Plate washer
3 Reservoir cap	10 Lever	17 Sealing washer - 11 off	24 Lower hose - 2 off
4 Countersunk screw - 4 off	11 Adjuster screw	18 Upper hose	25 Banjo bolt
5 Diaphragm	12 Locknut	19 Boot	26 Hose clamp
6 Gasket	13 Spring	20 Junction	27 Hose clamp
7 Clamp	14 Pivot bolt	21 Bolt	28 Front stop lamp switch

6 Removing and replacing the brake disc

1   It is unlikely that either disc will require attention until a considerable mileage has been covered, unless premature scoring of the disc has taken place thereby reducing braking efficiency. To remove each disc, first detach the front wheel as described in Chapter 4, Section 2.7 and 8. Each disc is bolted to the front wheel by six bolts, which are secured in pairs by a common tab washer. Bend back the tab washers and remove the bolts, to free the disc.

2   The brake disc can be checked for wear and for warpage whilst the front wheel is still in the machine. Using a micrometer, measure the thickness of the disc at the point of greatest wear. If the measurement is much less than the recommended service limit of 6.5 mm (0.26 in) the disc should be renewed. Check the warpage of the disc by setting up a suitable pointer close to the outer periphery of the disc and spinning the front wheel slowly. If the total warpage is more than 0.15 mm (0.006 in) the disc should be renewed. A warped disc, apart from reducing the braking efficiency, is likely to cause juddering during braking and will also cause the brake to bind when it is not in use.

7 Hydraulic brake hoses and pipes: examination

1   An external brake hose and pipe is used to transmit the hydraulic pressure to the caliper unit when the front brake or rear brake is applied. The brake hose is of the flexible type, fitted with an armoured surround. It is capable of withstanding pressures up to 350 kg/cm^2. The brake pipe attached to it is made from double steel tubing, zinc plated to give better corrosion resistance.

2   When the brake assembly is being overhauled, check the condition of both the hose and the pipe for signs of leakage or scuffing, if either has made rubbing contact with the machine whilst it is in motion. The union connections at either end must also be in good condition, with no stripped threads or damaged sealing washers. Check also the feed pipe from the rear brake master cylinder to the reservoir. This pipe is not subjected to pressure but may perish after a considerable length of time. It is a push fit on the unions and is retained by screw clips.

8 Bleeding the hydraulic system

1   As mentioned earlier, brake action is impaired or even rendered inoperative if air is introduced into the hydraulic system. This can occur if the seals leak, the reservoir is allowed to run dry or if the system is drained prior to the dismantling of any component part of the system. Even when the system is refilled with hydraulic fluid, air pockets will remain and because air will compress, the hydraulic action is lost.

2   Check the fluid content of the reservoir and fill almost to the top. Remember that hydraulic brake fluid is an excellent paint stripper, so beware of spillage, especially near the petrol tank.

3   Place a clean glass jar below the brake caliper unit and attach a clear plastic tube from the caliper bleed screw to the container. Place some clean hydraulic fluid in the container so that the pipe is always immersed below the surface of the fluid.

4   Unscrew the bleed screw one complete turn and pump the handlebar lever slowly. As the fluid is ejected from the bleed screw the level in the reservoir will fall. Take care that the level does not drop too low whilst the operation continues, otherwise air will re-enter the system, necessitating a fresh start.

5   Continue the pumping action with the lever until no further air bubbles emerge from the end of the plastic pipe. Hold the brake lever against the handlebars and tighten the caliper bleed screw. Remove the plastic tube AFTER the bleed screw is closed.

Where the front brakes are being bled, attach the pipe to the second caliper and repeat the sequence.

6   Check the brake action for sponginess, which usually denotes there is still air in the system. If the action is spongy, continue the bleeding operation in the same manner, until all traces of air are removed.

7   Bring the reservoir up to the correct level of fluid and replace the diaphragm, sealing gasket and cap. Check the entire system for leaks. Recheck the brake action.

8   Note that fluid from the container placed below the brake caliper unit whilst the system is bled, should not be reused, as it will have become aerated and may have absorbed moisture.

9 Front wheel bearings: examination and replacement

1   Access to the front wheel bearings may be made after removal of the wheel from the forks. Pull the speedometer gearbox out of the hub left-hand boss and remove the dust seal cover and wheel spacer from the hub right-hand side.

2   Lay the wheel on the ground with the disc side facing downward and with a special tool, in the form of a rod with a curved end, insert the curved end into the hole in the centre of the spacer separating the two wheel bearings. If the other end of the special tool is hit with a hammer, the right-hand bearing, bearing flange washer, and bearing spacer will be expelled from the hub.

6.1 Disc is secured by six bolts with lock plates

8.3 Place a pipe over the bleed nipple

3   Invert the wheel and drive out the left-hand bearing by inserting a drift of the appropriate size, through the hub. During the removal of either bearing it may be necessary to support the wheel across an open-ended box so that there is sufficient clearance for the bearing to be displaced completely from the hub.

4   Remove all the old grease from the hub and bearings, giving the latter a final wash in petrol. Check the bearings for signs of play or roughness when they are turned. If there is any doubt about the condition of a bearing, it should be renewed.

5   Before replacing the bearings, first pack the hub with new grease. Then drive the bearings back into position, not forgetting the distance piece that separates them. Take great care to ensure that the bearings enter the housings perfectly squarely otherwise the housing surface may be broached. Fit replacement oil seals and any dust covers or spacers that were also displaced during the original dismantling operation.

10 Rear wheel: examination, removal and renovation

1   Place the machine on the centre stand so that the rear wheel is raised clear of the ground. Check for rim alignment, spokes etc., as described for the front wheel in Section 2.

9.5a Do not omit bearing spacers or ...

9.1 Speedometer gearbox is a push fit in the hub

9.5b ... dust seals when replacing bearings

2   Removal of the rear wheel should be carried out as described in Chapter 4, Section 9.3-5.

11 Rear wheel bearings: examination and replacement

1   The procedure for the removal and examination of the rear wheel bearings is similar to that given for the front wheel bearings. A heavy dust cover and oil seals are fitted on both sides of the hub. Commence by drifting out the right-hand wheel bearing and bearing spacer. Two bearings placed side by side are fitted on the left of the hub. These should be drifted out together. The double sealed bearing should be fitted on the outside.

12 Rear disc brake: examination, pad inspection and overhaul

1   The disc and caliper assembly fitted to the rear of the machine are almost identical in design to the units fitted to the front. The procedure for inspection and repair is therefore similar.

2   The caliper may be detached complete with the mounting bracket after removal of the rear wheel and the bracket pivot nut. For normal purposes the caliper may be detached from the support bracket in a manner similar to that of the front caliper.

11.1a Rear wheel right-hand oil seal and ...

11.1b ... right-hand sealed bearing

11.1c Rear wheel has two bearings on left

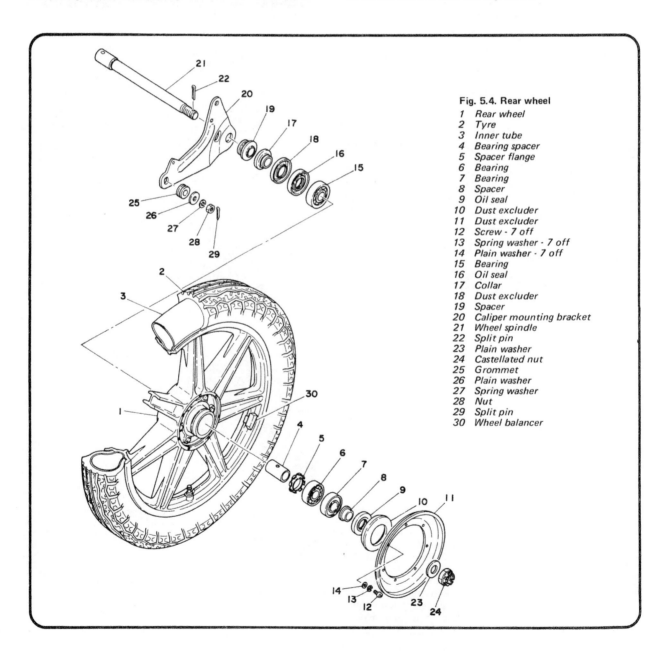

Fig. 5.4. Rear wheel

1   Rear wheel
2   Tyre
3   Inner tube
4   Bearing spacer
5   Spacer flange
6   Bearing
7   Bearing
8   Spacer
9   Oil seal
10 Dust excluder
11 Dust excluder
12 Screw - 7 off
13 Spring washer - 7 off
14 Plain washer - 7 off
15 Bearing
16 Oil seal
17 Collar
18 Dust excluder
19 Spacer
20 Caliper mounting bracket
21 Wheel spindle
22 Split pin
23 Plain washer
24 Castellated nut
25 Grommet
26 Plain washer
27 Spring washer
28 Nut
29 Split pin
30 Wheel balancer

13 Rear brake master cylinder: removal, examination and renovation

1   The rear brake master cylinder is mounted in board of the frame right-hand triangulation, and is so placed that the fluid level can be seen readily without the need to remove the side cover. The master cylinder is operated by a foot pedal via an adjustable push rod connected to the pedal shaft by a clevis pin and a split pin.

2   Drain the master cylinder and reservoir, using a similar technique to that described for the front brake master cylinder. The master cylinder reservoir is fitted with a triangular cap, secured by three screws.

3   Disconnect the hydraulic hose at the master cylinder by removing the banjo bolt. Take care not to drop any residual fluid on the paintwork. The master cylinder is retained on the frame lug by two bolts. After removal of the bolt, the cylinder unit may be lifted upwards so that the operating push rod leaves the cylinder. The master cylinder can now be lifted away from the machine.

4   Examination and dismantling of the rear brake master cylinder may be made by referring to the directions in Section 5 of this Chapter. Additionally, the reservoir should be flushed out with clean fluid before refitting.

13.1a Rear brake master cylinder is controlled by ...

5   After reassembly and replacement of the rear brake master cylinder components which may be made by reversing the dismantling procedure - bleed the rear brake system of air by referring to Section 7 of this Chapter.

14 Rear brake pedal height: adjustment

1   The pivot shaft upon which the rear brake pedal is mounted is splined to allow adjustment of the pedal height to suit individual requirements.

2   To adjust the height, loosen and remove the pinch bolt which passes into the rear of the pedal box. Draw the pedal off the splines and refit it at the required angle. Ideally the pedal should be fitted, so that it is positioned just below the rider's right foot, when the rider is seated normally. In this way the foot does not have to be lifted before the brake can be applied.

3   The upper limit of travel of the brake pedal may be adjusted by means of the bolt and locknut fitted to the pedal pivot mounting bracket. Care should be exercised when lowering the pedal by this method as movement imparted to the master cylinder piston may actuate the brake to a small degree. Adjustment of the pedal may necessitate readjustment of the rear stop lamp switch.

15 Tyres: removal and replacement

1   At some time or other the need will arise to remove and replace the tyres, either as the result of a puncture or because a renewal is required to offset wear. To the inexperienced, tyre changing represents a formidable task yet if a few simple rules are observed and the technique learned, the whole operation is surprisingly simple.

2   To remove the tyre from the wheel, first detach the wheel from the machine by following the procedure in Chapter 4, Sections 2.7 and 2.8 for the front wheel and Section 9.3-5 for the rear wheel. Deflate the tyre by removing the valve insert and when it is fully deflated, push the bead of the tyre away from the wheel rim on both sides so that the bead enters the centre well of the rim. Remove the locking cap and push the tyre valve into the tyre itself.

3   Insert a tyre lever close to the valve and lever the edge of the tyre over the outside of the wheel rim. Very little force should be necessary; if resistance is encountered it is probably due to the fact that the tyre beads have not entered the well of the wheel rim all the way round the tyre.

4   Once the tyre has been edged over the wheel rim, it is easy to work around the wheel rim so that the tyre is completely

13.1b ... rocking shaft and pushrod

free on one side. At this stage, the inner tube can be removed.

5   Working from the other side of the wheel ease the other edge of the tyre over the outside of the wheel rim furthest away. Continue to work around the rim until the tyre is free from the rim.

6   If a puncture has necessitated the removal of the tyre, reinflate the inner tube and immerse it in a bowl of water to trace the source of the leak. Mark its position and deflate the tube. Dry the tube and clean the area around the puncture with a petrol-soaked rag. When the surface has dried, apply rubber solution and allow this to dry before removing the backing from the patch and applying the patch to the surface.

7   It is best to use a patch of the self-vulcanising type, which will form a permanent repair. Note that it may be necessary to remove a protective covering from the top surface of the patch, after it has sealed in position. Inner tubes made from synthetic rubber may require a special type of patch and adhesive, if a satisfactory bond is to be achieved.

8   Before replacing the tyre, check the inside to make sure the agent that caused the puncture is not trapped. Check also the outside of the tyre, particularly the tread area, to make sure nothing is trapped that may cause a further puncture.

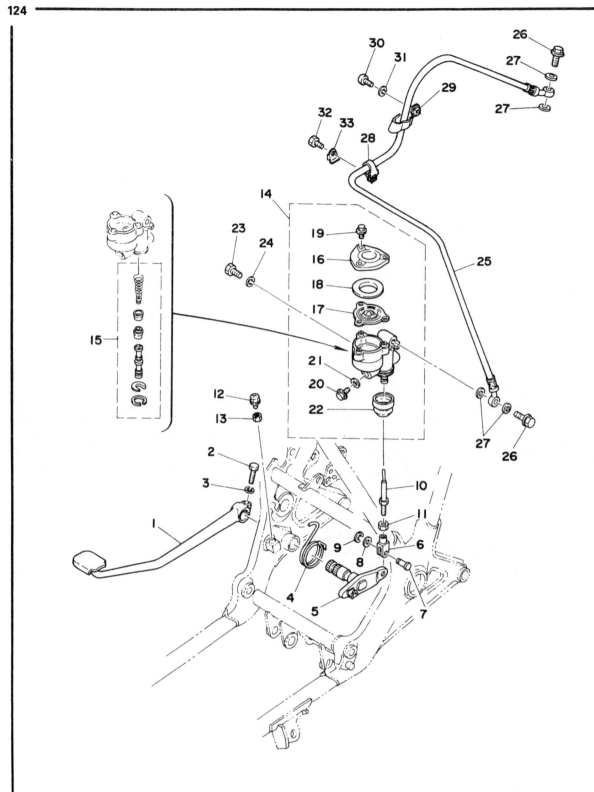

Fig. 5.5. Rear brake master cylinder

1	Brake pedal	9	'E' clip	17	Diaphragm	24	Spring washer - 2 off
2	Bolt	10	Pushrod	18	Gasket	25	Brake hose
3	Spring washer	11	Locknut	19	Bolt - 3 off	26	Banjo bolt - 2 off
4	Return spring	12	Adjuster screw	20	Plug	27	Sealing washer - 4 off
5	Brake pivot shaft	13	Locknut	21	Sealing washer	28	Clamp
6	Clevis fork	14	Master cylinder assembly	22	Boot	29	Clamp
7	Clevis pin	15	Piston kit	23	Bolt - 2 off	30	Screw
8	Plain washer	16	Reservoir cap				

Tyre changing sequence - tubed tyres

 Deflate tyre. After pushing tyre beads away from rim flanges push tyre bead into well of rim at point opposite valve. Insert tyre lever adjacent to valve and work bead over edge of rim.

Use two levers to work bead over edge of rim. Note use of rim protectors

 Remove inner tube from tyre

When first bead is clear, remove tyre as shown

 When fitting, partially inflate inner tube and insert in tyre

Work first bead over rim and feed valve through hole in rim. Partially screw on retaining nut to hold valve in place.

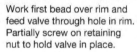

 Check that inner tube is positioned correctly and work second bead over rim using tyre levers. Start at a point opposite valve.

Work final area of bead over rim whilst pushing valve inwards to ensure that inner tube is not trapped

9 If the inner tube has been patched on a number of past occasions, or if there is a tear or large hole, it is preferable to discard it and fit a new one. Sudden deflation may cause an accident, particularly if it occurs with the front wheel.

10 To replace the tyre, inflate the inner tube sufficiently for it to assume a circular shape but only just. Then push it into the tyre so that it is enclosed completely. Lay the tyre on the wheel at an angle and insert the valve through the rim tape and the hole in the wheel rim. Attach the locking cap on the first few threads, sufficient to hold the valve captive in its correct location.

11 Starting at the point furthest from the valve, push the tyre bead over the edge of the wheel rim until it is located in the central well. Continue to work around the tyre in this fashion until the whole of one side of the tyre is on the rim. It may be necessary to use a tyre lever during the final stages.

12 Make sure there is no pull on the tyre valve and again commencing with the area furthest from the valve, ease the other bead of the tyre over the edge of the rim. Finish with the area close to the valve, pushing the valve up into the tyre until the locking cap touches the rim. This will ensure the inner tube is not trapped when the last section of the bead is edged over the rim with a tyre lever.

13 Check that the inner tube is not trapped at any point. Reinflate the inner tube, and check that the tyre is seating correctly around the wheel rim. There should be a thin rib moulded around the wall of the tyre on both sides, which should be equidistant from the wheel rim at all points. If the tyre is unevenly located on the rim, try bouncing the wheel when the tyre is at the recommended pressure. It is probable that one of the beads has not pulled clear of the centre well.

14 Always run the tyres at the recommended pressures and never under or over-inflate. The correct pressures for solo use are given in the Specifications Section of this Chapter.

15 Tyre replacement is aided by dusting the side walls, particularly in the vicinity of the beads, with a liberal coating of french chalk. Washing-up liquid can also be used to good effect.

16 Never replace the inner tube and tyre without the rim tape in position. If this precaution is overlooked there is good chance of the ends of the spoke nipples chafing the inner tube and causing a crop of punctures.

17 Never fit a tyre that has a damaged tread or side walls. Apart from the legal aspects, there is a very great risk of blow-out, which can have serious consequences on any two-wheeled vehicle.

18 Tyre valves rarely give trouble, but it is always advisable to check whether the valve itself is leaking before removing the tyre. Do not forget to fit the dust cap, which forms an effective second seal.

16 Valve cores and caps

1 Valve cores seldom give trouble, but do not last indefinitely. Dirt under the seating will cause a puzzling 'slow-puncture'. Check that they are not leaking by applying spittle to the end of the valve and watching for air bubbles.

2 A valve cap is a safety device, and should always be fitted. Apart from keeping dirt out of the valve, it provides a second seal in case of valve failure, and may prevent an accident resulting from sudden deflation.

17 Front wheel balancing

1 The front wheel should be statically balanced, complete with tyre. An out of balance wheel can produce dangerous wobbling at high speed.

2 Some tyres have a balance mark on the sidewall. This must be positioned adjacent to the valve. Even so, the wheel still requires balancing.

3 With the front wheel clear of the ground, spin the wheel several times. Each time, it will probably come to rest in the same position. Balance weights should be attached diametrically opposite the heavy spot, until the wheel will not come to rest in any set position, when spun.

4 Machines fitted with cast aluminium wheels require special balancing weights which are designed to clip onto the centre rim flange, much in the way that weights are affixed to car wheels. When fitting these weights, take care not to affix any weight nearer than 40 mm (1.54 in) to the radial centre line of any spoke. Refer to the accompanying diagram.

5 It is possible to have a wheel dynamically balanced at some dealers. This requires its removal.

6 There is no need to balance the rear wheel under normal road conditions, although any tyre balance mark should be aligned with the valve.

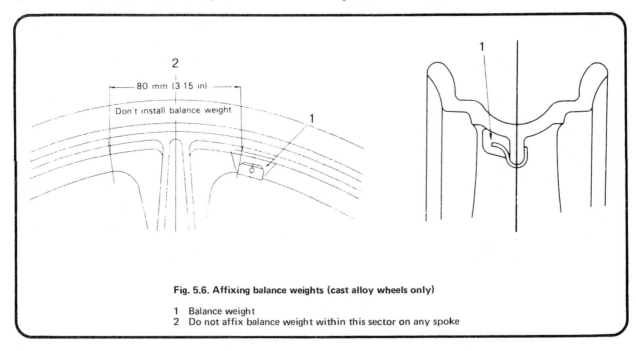

Fig. 5.6. Affixing balance weights (cast alloy wheels only)

1 Balance weight
2 Do not affix balance weight within this sector on any spoke

18 Fault diagnosis: wheels, brakes and tyres

Symptom	Cause	Remedy
Handlebars oscillate at low speeds	Buckle or flat in wheel rim, most probably front wheel (spoked wheels only)	Check rim alignment by spinning wheel. Correct by retensioning spokes or rebuilding on new rim.
	Tyre not straight on rim	Check tyre alignment.
Machine lacks power and accelerates poorly	Brakes binding	Check brake pads and whether piston(s) sticking.
Brakes feel spongy	Air in hydraulic systems	Bleed brakes.
Brake pull-off sluggish	Sticking pistons in brake caliper	Overhaul caliper unit.

Chapter 6 Electrical system

Contents

Specifications

Battery

Make	Yuasa
Type	YB 14L
Voltage	12 volts
Capacity	14 ah
Earth	Negative

Alternator

Make	Hitachi
Model	Ld 120 - 02
Output	280 watts
Field coil resistance (at 20°C)	4.04 ± 0.4 ohms
Stator coil resistance (at 20°C)	0.48 ± 0.05 ohms

Starter motor

Make	Mitsuba
Model	SM - 224C
Brush length	12.5 mm (0.5 in)
Wear limit	5.5 mm (0.22 in)
Commutator undercut	0.5 - 0.8 mm (0.02 - 0.03 in)

*Bulbs

Headlamp	50/40 watt
Tail/stop lamp	8/27 watt x 2
License plate lamp (USA only)	8 watt
Speedometer	3.4 watt
Tachometer	3.4 watt
Neutral indicator	3.4 watt
High beam indicator	3.4 watt
Flashing indicators	27 watt x 4
Flasher indicator bulb	3.4 watt
Oil pressure warning	3.4 watt

All bulbs rated at 12 volts.

1 General description

1 The Yamaha XS 750 model is fitted with a 12 volt electrical system. The circuit comprises a crankshaft-driven alternator, the output of which is controlled by a voltage regulator linked with the field coil windings. Because the output from the alternator is AC, a rectifier is included in the circuit to convert to DC in order to maintain the charge of the 12 volt, 14 amp hour battery. The alternator has a built-up rotor attached to the extreme right-hand end of the crankshaft, rotating within a stator contained in the alternator cover casing. The rotor is not permanently magnetised, but is energised electro-magnetically by a fixed field coil, which is fitted in the alternator cover in a manner similar to that of the stator coil.

2 Crankshaft alternator: checking the output

1 The output from the alternator mounted on the end of the crankshaft can be checked only with specialised test equipment of the multimeter type. It is unlikely that the average owner/rider will have access to this equipment or instruction in its use. In consequence, if the performance of the alternator is in any way suspect, it should be checked by a Yamaha agent or an auto-electrical specialist.

2 If a multimeter is available, a general check on the alternator may be carried out as follows. Connect a DC voltmeter across the two battery terminals. Start the engine and increase the speed to 2000 rpm. The generated voltage should be 14 - 15 volts. If the voltage is within that range it may be assumed that the system is functioning correctly. Where small variations are evident, the voltage regulator should be checked and adjusted, as described in the following Section. A marked reduction in output may be a result of damaged windings in the stator coils or field coils. These may be checked for continuity and resistance as follows.

3 Remove the four socket screws holding the alternator cover in place. Lift the cover away, complete with the field coil and stator unit. Detach the left-hand side cover and disconnect the block connector with the three white wires and the block connector with the green wire and black wire. Check that no visible signs of damage to the stator coils or field coil is apparent. If all the windings appear in good order connect an ohmmeter across the wires as follows and check the resistance.

Field Coil
Green - Black   4.04 ohms ± 10% at 20°C (68°F)

*Stator Coils
  W1-W2
  W2-W3   0.48 ohms ± 10% at 20°C (68°F)
  W3-W1

* See accompanying diagram.

If the figure obtained for any one test is outside the range given, check that the wires between the coil and the block connector have not fractured. A damaged coil cannot be repaired, it must be replaced as a complete unit.

4 If the alternator is found to be in good condition the fault should be traced by checking the voltage regulator followed by the silicon rectifier.

3 Voltage regulator: examination and adjustment

1 The voltage regulator is fitted to the right-hand side of the machine adjacent to the battery carrier box. If the battery is continually overcharged or discharged, the regulator may be defective. Symptoms of overcharging include a battery that requires continual topping up of the electrolyte and lamp bulbs which burn out at high engine speeds.

2 If the regulator unit is suspect, a preliminary check may be made as follows:
    Ensure that the battery has a charge of not less than 12 volts. If the battery is undercharged or defective it is likely that the regulator unit will not function correctly. Remove the left-hand side cover and disconnect the block connector (5 wires, red, black and 3 white) from the rectifier which is secured to the underside of the battery box. Detach the red wire at the rectifier side of connector and then reconnect the block. Connect the positive lead of a 0-20 volt DC voltmeter to the red wire from the rectifier and connect the negative lead to a good earth point. Start the engine. The voltmeter should indicate a reading of 14.5 - 15 volts. If the voltage fluctuates by more than 0.5 volts, or if the stable voltage is outside the range, the voltage regulator is in need of adjustment.

3 Disconnect the battery and remove the voltage regulator from the frame. Detach the regulator cover and inspect the condition of the contact points. The points may be cleaned using very fine emery paper (No. 500 or 600) followed by methylated spirits. Check the core gap and points gap using a feeler gauge, making adjustments where required.

Core gap   0.6 - 1.0 mm (0.0236 - 0.0393 in)
Points gap   0.3 - 0.4 mm (0.0118 - 0.0157 in)

4 After adjustment, refit the regulator (without the cover) and reconnect the battery and regulator leads. Carry out the test described in paragraph 2. If the voltage is incorrect, stop the engine and loosen the locknut on the adjuster screw which passes through the frame of the regulator and abuts against the spring steel moving contact arm blade. Turn the screw 1/8 of a turn. Start the engine and recheck the voltage. Repeat this procedure until the correct voltage is given, then tighten the locknut. It will be noted that as the engine speed increases, the moving contact will move from the low fixed point to the high fixed point. As this occurs there will be a small fluctuation in voltage. The voltage fluctuation must be 0.5 volts. If the fluctuation is greater or lesser than 0.5 volts, the core gap must be adjusted once more.

4 Silicon rectifier: location and checking

1 The silicon rectifier fitted to all models converts the AC current produced by the alternator to DC so that it can be used to charge the battery.

2 The rectifier is fixed to the underside of the battery carrier box and may be reached by removal of the left-hand frame side cover. The rectifier cannot be repaired and if it is damaged it will have to be renewed. Damage can occur to the rectifier if the machine is run without a battery or if the battery leads are reversed.

3 The rectifier can be checked whilst still in position using an ohmmeter and referring to the following table. Disconnect the rectifier block connector and check the continuity of the individual diodes. Connect the positive lead of the ohmmeter, to the first wire colour in the table and the negative lead to the second wire colour. Continue the check with each pair of wires. In all cases there should be continuity. Repeat the process with the ohmmeter leads transposed. In all cases there should be no continuity.

D1	Red (+)	U (white)
D2	Red (+)	V (white)
D3	Red (+)	W (white)
D4	U (white)	Black (−)
D5	V (white)	Black (−)
D6	W (white)	Black (−)

If a discrepancy is evident, the complete rectifier unit must be renewed.

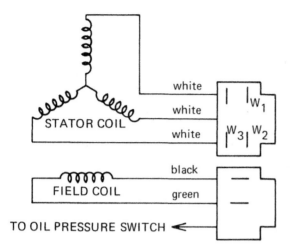

Fig. 6.1. Alternator coil resistance test

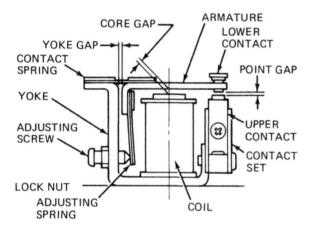

Fig. 6.2. Voltage regulator - schematic diagram

5 Battery: examination and maintenance

1 A Yuasa YB 14L battery is fitted as standard. This battery is a lead-acid type and has a capacity of 14 amp hours.
2 The transparent plastic case of the battery permits the upper and lower levels of the electrolyte to be observed when the battery is lifted from its housing below the dualseat. Maintenance is normally limited to keeping the electrolyte level between the prescribed upper and lower limits and by making sure the vent pipe is not blocked. The lead plates and their separators can be seen through the transparent case, a further guide to the general condition of the battery.
3 Unless acid is spilt, as may occur if the machine falls over, the electrolyte should always be topped up with distilled water, to restore the correct level. If acid is spilt on any of the machine, it should be neutralised with an alkali such as washing soda and washed away with plenty of water, otherwise serious corrosion will occur. Top up with sulphuric acid of the correct specific gravity (1.260 - 1.280) only when spillage has occurred. Check that the vent pipe is well clear of the frame tubes or any of the other cycle parts, for obvious reasons.

6 Battery: charging procedure

1 The normal charging rate for the 14 amp hour battery is 1.4 amps. A more rapid charge, not exceeding 2 amps can be given in an emergency. The higher charge rate should, if possible, be avoided since it will shorten the working life of the battery.
2 Make sure that the battery charger connections are correct, red to positive and black to negative. It is preferable to remove the battery from the machine whilst it is being charged and to remove the vent plug from each cell. When the battery is re-connected to the machine, the black lead must be connected to the negative terminal and the red lead to positive. This is most important, as the machine has a negative earth system. If the terminals are inadvertently reversed, the electrical system will be damaged permanently. The rectifier will be destroyed by a reversal of the current flow.

7 Fuse: location and replacement

1 A bank of fuses is contained within a small plastic box located near the regulator and sharing the same mounting bracket. The box contains four 10A fuses and two 20A fuses, of which one of each type is spare.
2 Before replacing a fuse that has blown, check that no obvious short circuit has occurred, otherwise the replacement fuse will blow immediately it is inserted. It is always wise to check the electrical circuit thoroughly, to trace the fault and eliminate it.
3 When a fuse blows while the machine is running and no spare is available, a 'get you home' remedy is to remove the blown fuse and wrap it in silver paper before replacing it in the fuseholder. The silver paper will restore the electrical continuity by bridging the broken fuse wire. This expedient should NEVER be used if there is evidence of a short circuit or other major electrical fault, otherwise more serious damage will be caused. Replace the 'doctored' fuse at the earliest possible opportunity, to restore full circuit protection.

7.1 Fuses are contained within a box on rear mudguard

8 Starter motor: removal, examination and replacement

1 An electric starter motor, operated from a small push-button on the right-hand side of the handlebars, provides an alternative and more convenient method of starting the engine, without having to use the kickstart. The starter motor is mounted within a compartment at the rear of the cylinder block, closed by an oblong plastic cover. Current is supplied from the battery via a heavy duty solenoid switch and a cable capable of carrying the very high current demanded by the starter motor on the initial start-up.
2 The starter motor drives a free running clutch via an idler pinion. The clutch also incorporates the oil pump drive pinion and shaft. The clutch ensures the starter motor drive is

disconnected from the pump drive immediately the engine starts. It operates on the centrifugal principle; spring loaded rollers take up the drive until the centrifugal force of the rotating engine overcomes their resistance and the drive is automatically disconnected.

3   To remove the starter motor from the engine unit, first it will be necessary to detach the carburettors to gain access as described in Chapter 2, Section 5. Disconnect the battery leads to isolate the electrical system and then remove the starter motor cover, retained by two screws. Detach the heavy cable from the terminal on the motor body.

4   The starter motor is secured to the crankcase by two bolts which pass through the left-hand end of the motor casing. When these bolts are withdrawn, the motor can be prised out of position and lifted out of its compartment. The starter motor is a very tight fit. Take care not to lever the upper portion of the motor boss against the top of the gear casing inner wall. **The casing may fracture if overstressed.** The parts of the starter motor most likely to require attention are the brushes. The end cover is retained by the two long screws which pass through the lugs cast on both end pieces. If the screws are withdrawn, the end cover can be lifted away and the brush gear exposed.

5   Lift up the spring clips which bear on the end of each brush and remove the brushes from their holders. Each brush should

have a length of 12.5 mm (0.5 in). The minimum allowable brush length is 5.5 mm (0.22 in). If the brush is shorter it must be renewed.

6   Before the brushes are replaced, make sure that the commutator is clean. The commutator is the copper segments on which the brushes bear. Clean the commutator with a strip of glass paper. Never use emery cloth or 'wet-and-dry' as the small abrasive fragments may embed themselves in the soft copper of the commutator and cause excessive wear of the brushes. Finish off the commutator with metal polish to give a smooth surface and finally wipe the segments over with a methylated spirits soaked rag to ensure a grease free surface. Check that the mica insulators, which lie between the segments of the commutator, are undercut. The standard groove depth is 0.5 - 0.8 mm (0.02 - 0.03 in), but if the average groove depth is less than this the armature should be renewed or returned to a Yamaha dealer for re-cutting.

7   Replace the brushes in their holders and check that they slide quite freely. Make sure the brushes are replaced in their original positions because they will have worn to the profile of the commutator. Replace and tighten the end cover, then replace the starter motor and cable in the housing, tighten down and remake the electrical connection to the solenoid switch. Check that the starter motor functions correctly before replacing the compartment cover and sealing gasket.

8.2a Starter clutch consists of spring loaded rollers in a drum

8.2b The clutch unit rotates on the oil pump idler shaft

8.5 Lift up springs to allow brush removal

8.6 Check commutator for wear and mica undercut

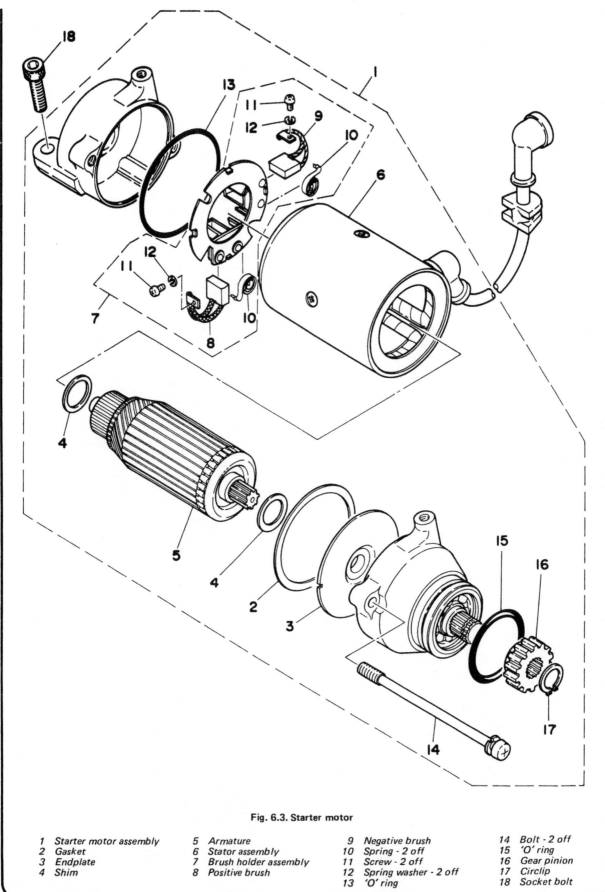

Fig. 6.3. Starter motor

1	Starter motor assembly	5	Armature	9	Negative brush	14 Bolt - 2 off
2	Gasket	6	Stator assembly	10	Spring - 2 off	15 'O' ring
3	Endplate	7	Brush holder assembly	11	Screw - 2 off	16 Gear pinion
4	Shim	8	Positive brush	12	Spring washer - 2 off	17 Circlip
				13	'O' ring	18 Socket bolt

8.7 End cover must align with index marks

9 Starter solenoid switch: function and location

1   The starter motor switch is designed to work on the electro-magnetic principle. When the starter motor button is depressed, current from the battery passes through windings in the switch solenoid and generates an electro-magnetic force which causes a set of contact points to close. Immediately the points close, the starter motor is energised and a very heavy current is drawn from the battery.
2   This arrangement is used for at least two reasons. Firstly, the starter motor current is drawn only when the button is depressed and is cut off again when pressure on the button is released. This ensures minimum drainage on the battery. Secondly, if the battery is in a low state of charge, there will not be sufficient current to cause the solenoid contacts to close. In consequence, it is not possible to place an excessive drain on the battery which, in some circumstances, can cause the plates to overheat and shed their coatings. If the starter will not operate, first suspect a discharged battery. This can be checked by trying the horn or switching on the lights. If this check shows the battery to be in good shape, suspect the starter switch which should come into action with a pronounced click. It is located behind the left-hand side panel and can be identified by the heavy duty starter cable connected to it. It is not possible to effect a satisfactory repair if the switch malfunctions; it must be renewed.

10 Headlamp: replacing the bulbs and adjusting beam height

1   In order to gain access to the headlamp bulbs it is necessary to first remove the rim, complete with the reflector and headlamp glass. The rim is retained by one or two screws (depending on the model) which pass through the headlamp shell just below the two headlamp mounting bolts.
2   The headlamp unit fitted to the US model is of the sealed beam type and therefore if either of the headlamp filaments blow the complete unit should be replaced. On other models the headlamp bulb is a bayonet fit in the rear of the reflector. A reflector that accepts a pilot bulb is fitted to all models delivered to countries or states where parking lights are a statutory requirement. The pilot bulb is held in the bulb holder by a bayonet fixing.
3   Beam height on all models is effected by tilting the headlamp shell about the axis of the headlamp/front indicator lamp, mounting bolts. To make the adjustment loosen the bolt which passes through the adjuster bracket. The headlamp can then be moved within the confines of the elongated hole in the bracket. Horizontal alignment of the beam can be adjusted by altering

the position of the screw which passes through the headlamp rim. The screw is fitted at the 9 o'clock position when viewed from the front of the machine. Turning the screw in a clockwise direction will move the beam direction over to the left-hand side.
4   In the UK, regulations stipulate that the headlamp must be arranged so that the light will not dazzle a person standing at a distance greater than 25 feet from the lamp, whose eye level is not less than 3 feet 6 inches above that plane. It is easy to approximate this setting by placing the machine 25 feet away from a wall, on a level road, and setting the beam height so that it is concentrated at the same height as the distance of the centre of the headlamp from the ground. The rider must be seated normally during this operation and also the pillion passenger, if one is carried regularly.

11 Stop and tail lamp: replacing the bulbs

1   On Yamaha XS 750 models with European specifications the tail lamp is fitted with two twin filament bulbs of 12 volt, 5/21w rating, to illuminate the rear number plate and rear of the machine, and to give visual warning when the rear brake is applied. To gain access to the bulbs remove the plastic lens cover, which is retained by two long screws. Check that the gasket between the lens cover and the main body of the lamp is in good condition.
2   Each bulb has a bayonet fitting and has staggered pins to prevent the bulb contacts from being reversed.
3   On models originally supplied to the US, a similar rear lamp with two 32/3 cp bulbs is fitted. In addition, a separate small lamp is provided, mounted below the main lamp, which illuminates the rear number plate. The lamp is fitted with an 8w bulb.
4   If the tail lamp bulbs keep blowing, suspect either vibration of the rear mudguard or more probably, an intermittent earth connection.

12 Flashing indicator lamps

1   The forward facing indicator lamps are connected to 'stalks' that replace the bolts on which the headlamp shell would normally be mounted. The stalks are hollow and have threaded ends so that they can be locked in position from the inside of the headlamp shell or to the fork lugs that carry the headlamp. The rear facing lamps are mounted on similar, shorter stalks, at a point immediately to the rear of the dualseat.
2   In each case, access to the bulb is gained by removing the plastic lens cover, which is retained by two screws. Bayonet fitting bulbs of the single filament type are used, each with a 12 volt 27w rating (UK) or 12 volt 32 cp (USA).

9.1 Starter motor solenoid - fitted with heavy cables

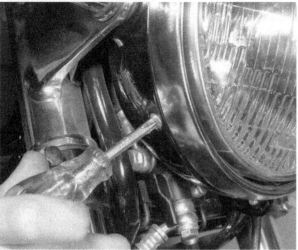

10.2a Headlamp rim retained by one or two screws

10.2b Socket is a push fit at rear of reflector

11.1 Tail lamp is fitted with two bulbs

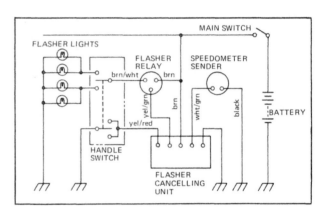

Fig. 6.4 Self cancelling indicator circuit diagram

13 Flasher unit: location and replacement

1   The flasher relay unit is located behind the right-hand side cover and is supported on anti-vibration mounting made of rubber.
2   If the flasher unit is functioning correctly, a series of audible clicks will be heard when the indicator lamps are in action. If the unit malfunctions and all the bulbs are in working order, the usual symptom is one initial flash before the unit goes dead; it will be necessary to replace the unit complete if the fault cannot be attributed to any other cause.
3   Take great care when handling the unit as it is easily damaged if dropped.

14 Self cancelling flashing indicator system: general description and testing

1   The purpose of this system is to turn off the turn signal automatically after a period of time or distance. At a very low speed the signal will cancel after a distance of 142 yards (130 metres) has been covered. At high speeds the signal will cancel after a time of 10 seconds has elapsed. When travelling in

the lower range of speeds the signal will cancel after a combination of both time and distance.
2   If the handlebar switch lever is moved to the left or right turn positions it will return directly to the OFF position but the signal will continue to function until automatically cancelled electrically. By pushing the same lever in, the signal may be cancelled manually.
3   Should the system fail to operate, carry out the following test procedure. Pull off the 6-pin connector from the self cancelling unit and operate the handlebar switch. If the signals operate normally in the L, R and OFF positions then the bulbs, lighting circuit, handlebar switch light circuit and flasher unit are all in good operating condition.
4   If the previous check is satisfactory, then the flasher cancelling unit, the handlebar switch reset circuit or the speedometer sensor circuit may be faulty. These components may be tested by carrying out the following test procedures.
5   Detach the 6-pin connector from the flasher cancelling unit and connect an ohmmeter with a 0 - 100 ohm range across the white/green and the black leads on the harness side. Turn the speedometer shaft. The ohmmeter needle should swing back and forth four times between zero and infinity on the scale, indicating that the speedometer sensor circuit is in good condition. If no needle deflection is apparent then the sensor or wire harness may be inoperative.

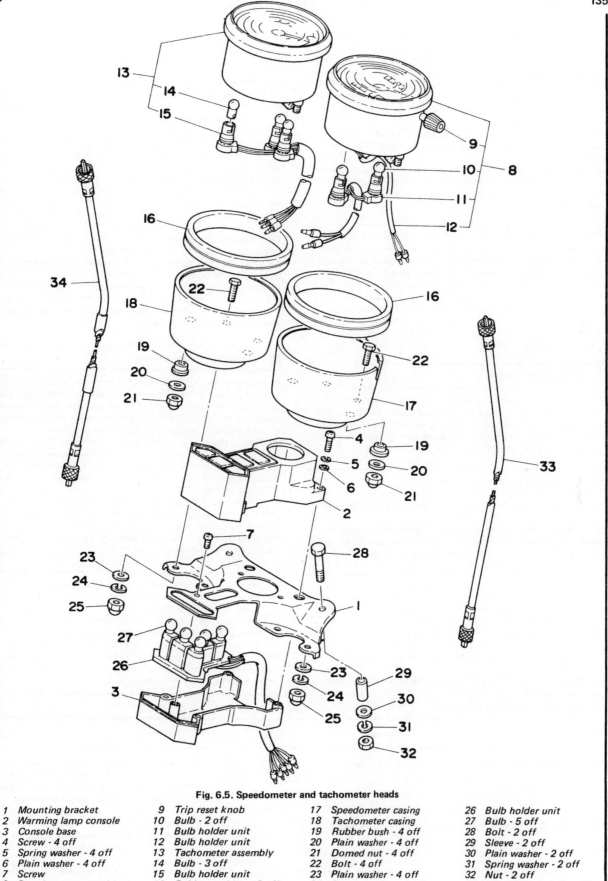

Fig. 6.5. Speedometer and tachometer heads

1 Mounting bracket	9 Trip reset knob	17 Speedometer casing	26 Bulb holder unit
2 Warming lamp console	10 Bulb - 2 off	18 Tachometer casing	27 Bulb - 5 off
3 Console base	11 Bulb holder unit	19 Rubber bush - 4 off	28 Bolt - 2 off
4 Screw - 4 off	12 Bulb holder unit	20 Plain washer - 4 off	29 Sleeve - 2 off
5 Spring washer - 4 off	13 Tachometer assembly	21 Domed nut - 4 off	30 Plain washer - 2 off
6 Plain washer - 4 off	14 Bulb - 3 off	22 Bolt - 4 off	31 Spring washer - 2 off
7 Screw	15 Bulb holder unit	23 Plain washer - 4 off	32 Nut - 2 off
8 Speedometer assembly	16 Rubber cushion - 2 off	24 Spring washer - 4 off	33 Speedometer drive cable
		25 Domed nut - 4 off	34 Tachometer drive cable

6 With the 6-pin connector detached from the flasher cancelling unit, check for continuity between the yellow/red lead on the harness side and the frame. With the handlebar switch set to the L or R position there should be a zero reading on the ohmmeter. With the switch set to the OFF position the needle should deflect to infinity. If the ohmmeter readings are not as stated, check the handlebar switch circuit and the wire harness for continuity.

7 If after completing the above checks the flasher cancelling system is still inoperative, then the cancelling unit must be renewed.

8 Should the flashing indicators operate only when the handlebar switch lever is moved to the L or R positions and turn off immediately the switch lever returns to the centre position, renew the cancelling unit.

15 Speedometer and tachometer heads: replacement of bulbs

1 Bulbs fitted to each instrument illuminate the dials during the hours of darkness when the headlamp is switched on. All bulbs fitted to either instrument head have the same type of bulb holder which is a push fit in the instrument base.

2 Access to the bulbs and holders is gained by removing the nuts and washers which secure the rubber mounted instruments to their protective outer cases. Disconnect the drive cables and lift each instrument up and pull out the bulb holders.

16 Indicator panel lamps

1 An indicator lamp panel which holds five warning bulbs is fitted between the speedometer and tachometer heads. To gain access to the bulbs remove the four screws which pass through the upper cover and lift the cover away, over the ignition switch barrel. Each bulb is a screw fit in the one piece holder.

17 Horn: location and examination

1 The horn is suspended from a flexible steel strip bolted to the frame below the petrol tank.

2 The horn has no external means of adjustment. If it malfunctions, it must be renewed; it is a statutory requirement that the machine must be fitted with a horn in working order.

18 Ignition switch: removal and replacement

1 The combined ignition and lighting master switch is mounted in the warning light panel mounting plate.

2 If the switch proves defective it may be removed after detaching the warning light cover and unscrewing the two mounting screws. Disconnect the two wiring sockets connecting the switch to the loom.

3 Reassembly of the switch can be made in the reverse pro-

cedure as described for dismantling. Repair is rarely practicable. It is preferable to purchase a new switch unit, which will probably necessitate the use of a different key.

19 Stop lamp switch: adjustment

1 All models have a stop lamp switch fitted to operate in conjunction with the rear brake pedal. The switch is located immediately to the rear of the crankcase, on the right-hand side of the machine. It has a threaded body giving a range of adjustment.

2 If the stop lamp is late in operating, slacken the locknuts and turn the body of the lamp in an anticlockwise direction so that the switch rises from the bracket to which it is attached. When the adjustment seems near correct, tighten the locknuts and test.

3 If the lamp operates too early, the locknuts should be slackened and the switch body turned clockwise so that it is lowered in relation to the mounting bracket.

4 As a guide, the light should operate after the brake pedal has been depressed by about 2 cm (¾ inch).

5 A stop lamp switch is also incorporated in the front brake system. The mechanical switch is a push fit in the handlebar lever stock. If the switch malfunctions, repair is impracticable. The switch should be renewed.

20 Handlebar switches: general

1 Generally speaking, the switches give little trouble, but if necessary they can be dismantled by separating the halves which form a split clamp around the handlebars. Note that the machine cannot be started until the ignition cut-out on the right-hand end of the handlebars is turned to the central 'ON' position.

2 Always disconnect the battery before removing any of the switches, to prevent the possibility of a short circuit. Most troubles are caused by dirty contacts, but in the event of the breakage of some internal part, it will be necessary to renew the complete switch.

3 Because the internal components of each switch are very small, and therefore difficult to dismantle and reassemble, it is suggested a special electrical contact cleaner be used to clean corroded contacts. This can be sprayed into each switch, without the need for dismantling.

21 Neutral indicator switch: location and removal

1 A switch is incorporated in the gearbox which indicates via a small light in the warning console when neutral gear has been selected. The switch is screwed into the base of the gearbox. In the event of failure the switch may be unscrewed without draining the transmission oil. Disconnect the switch lead by unscrewing the central cross-head screw. Note that the switch is prone to shear if it is overtightened.

22 Fault diagnosis: electrical system

Symptom	Cause	Remedy
Complete electrical failure	Blown fuse	Check wiring and electrical components for short circuit before fitting a new fuse. Check battery connections, also whether connections show signs of corrosion.
Dim lights, horn inoperative	Discharged battery	Recharge battery with battery charger and check whether alternator is giving correct output (electrical specialist).
Constantly 'blowing' bulbs	Vibration, poor earth connection	Check whether bulb holders are secured correctly. Check earth return or connections to frame.

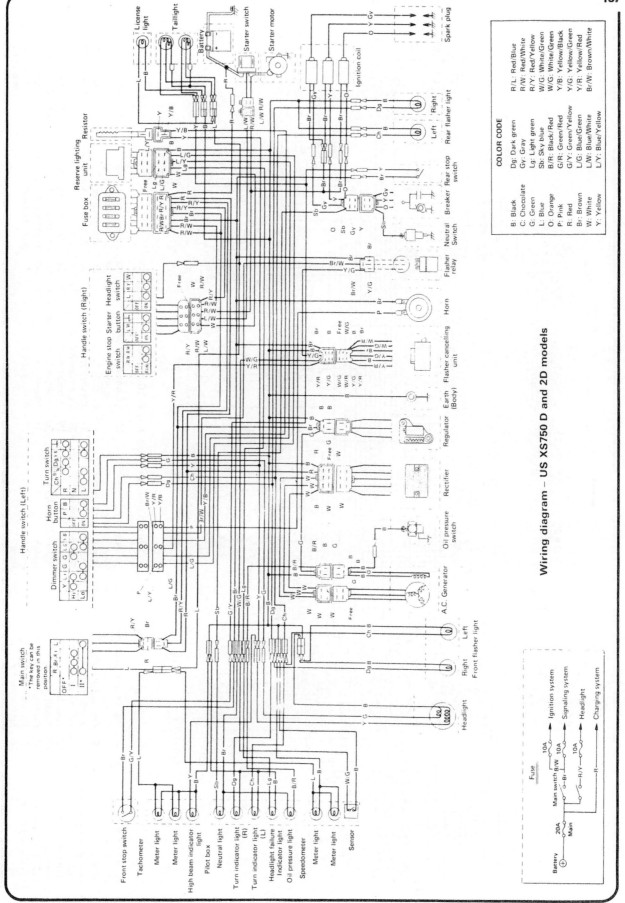

Wiring diagram – US XS750 D and 2D models

COLOR CODE

B: Black	Dg: Dark green	R/L: Red/Blue
C: Chocolate	Gy: Gray	R/W: Red/White
G: Green	Lg: Light green	R/Y: Red/Yellow
L: Blue	Sb: Sky blue	W/G: White/Green
O: Orange	B/R: Black/Red	W/G: White/Green
P: Pink	G/R: Green/Red	Y/B: Yellow/Black
R: Red	G/Y: Green/Yellow	Y/G: Yellow/Green
Br: Brown	L/G: Blue/Green	Y/R: Yellow/Red
W: White	L/W: Blue/White	Br/W: Brown/White
Y: Yellow	L/Y: Blue/Yellow	

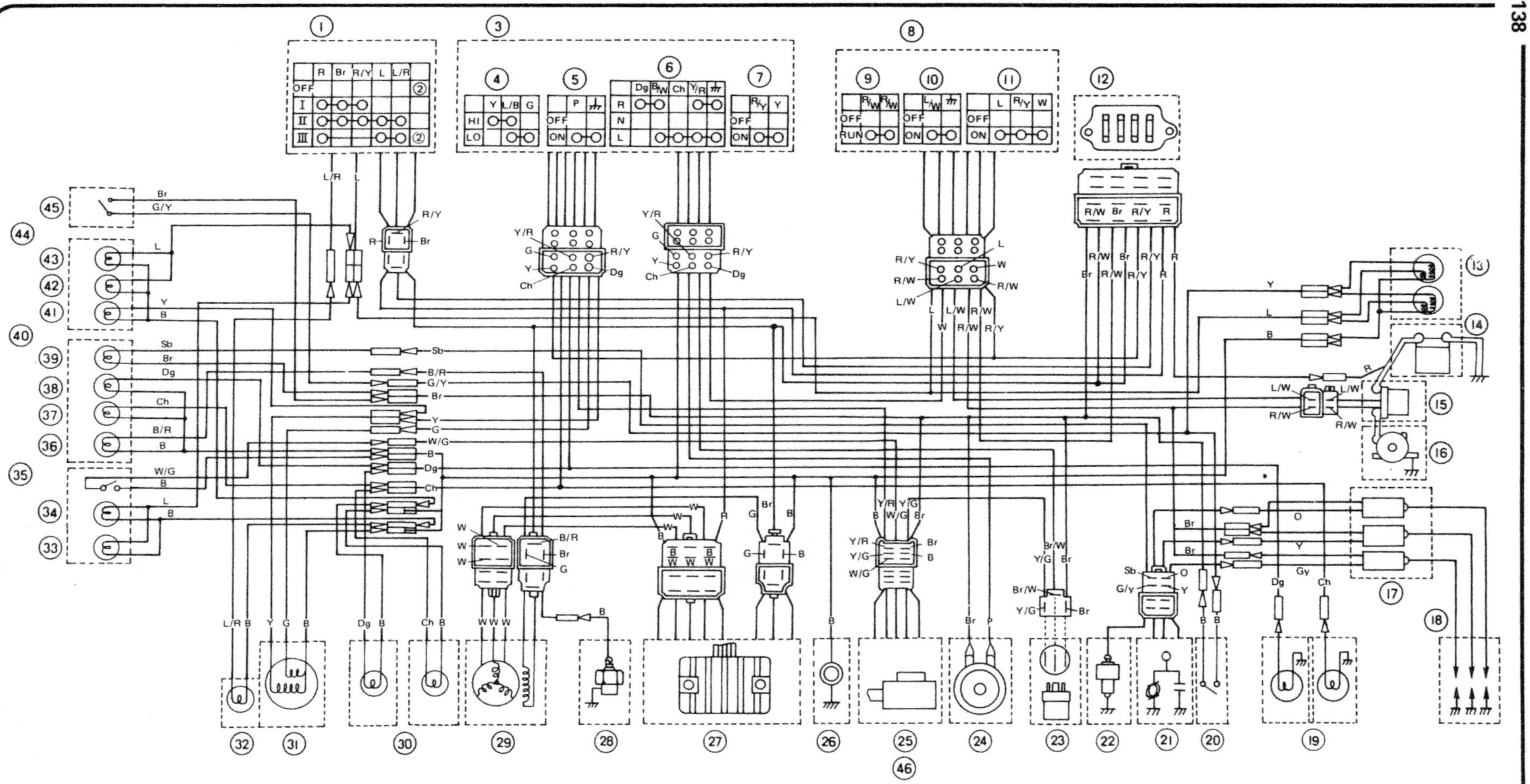

Wiring diagram – UK XS750 2D model

1 Main switch
2 Parking switch position
3 Left-hand handlebar switch
4 Dimmer switch
5 Horn button
6 Indicator switch
7 Headlamp flasher
8 Right-hand handlebar switch
9 Engine kill button
10 Starter button
11 Headlamp switch
12 Fuse box
13 Tail/stop lamp
14 Battery
15 Starter solenoid
16 Starter motor
17 Ignition coil
18 Spark plug
19 Rear indicator lamps
20 Rear stop lamp switch
21 Contact breaker assembly
22 Neutral indicator switch
23 Indicator relay
24 Horn
25 Indicator self cancelling unit
26 Earth
27 Regulator/rectifier
28 Oil pressure switch
29 Alternator
30 Front indicator lamps
31 Headlamp
32 Parking lamp
33 Speedometer lamp
34 Speedometer lamp
35 Speedometer
36 Oil pressure warning lamp
37 Left-hand indicator warning lamp
38 Right-hand indicator warning lamp
39 Neutral indicator lamp
40 Pilot lamp box
41 High beam indicator lamp
42 Tachometer lamp
43 Tachometer lamp
44 Tachometer
45 Front brake stop lamp switch

The 1979 Yamaha XS750 E model

Chapter 7 The XS 750 E, SE, F and SF models

Contents

Specifications

Specifications for the XS750 E, XS750 SE, XS750 F and XS750 SF models are as stated in the main text of this manual except where listed below:

Specifications relating to Chapter 1

Engine

Compression ratio:

XS750 E (UK)	8.5 : 1
All others	9.2 : 1

Valve guide/stem clearance:

Inlet	0.010 — 0.040 mm (0.0004 — 0.0016 in)
Service limit	0.10 mm (0.004 in)
Exhaust	0.025 — 0.055 mm (0.0010 — 0.0022 in)
Service limit	0.12 mm (0.005 in)

Valve guide ID:

Inlet and exhaust	7.0 — 7.015 mm (0.2756 — 0.2762 in)
Service limit	7.10 mm (0.2795 in)

Valve clearances – UK XS750 SE only (engine cold)

Inlet	0.11-0.15 mm (0.004-0.006in)
Exhaust	0.21-0.25 mm (0.008-0.010in)

Clutch

Spring free length service limit	41.8 mm (1.65 in)
XS750 E model only	41.5 mm (1.63 in)

Gearbox

Gear ratios, overall:

1st	14.697 : 1 (32/13)
2nd	9.483 : 1 (27/17)
3rd	7.762 : 1 (26/20)
4th	6.539 : 1 (23/21)
5th	5.711 : 1 (22/23)

Specifications relating to Chapter 2

Carburettors

Make	Mikuni
Type:	
XS750 SE (UK)	BS 34 – 11 2K100
XS750 SE (US) and XS750 SE and SF EPA models ...	BS 34 2G2 – 00
All other models	BS 34 2F3 – 00
Main jet	130
Jet needle	5H x 6
Needle clip position:	
XS750 SE (UK)	4
All other models	3
Needle jet	Y0
Starter jet	30
Pilot jet	42.5
Float height:	
XS750 E, SE and SF EPA models. XS750 F (US)	25.7 ± 1.0 mm (1.012 ± 0.040 in)
All other models	25.7 ± 2.5 mm (1.012 ± 0.10 in)
Pilot screw	Preset
Main air jet:	
XS750 E (US)	170
All other models	140
Pilot air jet	180
Throttle valve	135

Specifications relating to Chapter 3

Ignitor unit

Make	Hitachi
Type	TID 03 – 01
Capacity:	
XS750 E (UK) and XS750 SE (UK)	5.5 amps or less
All other models	4.0 amps or less
Frequency	One ignition/cycle (each cylinder)

Ignition timing

Retarded:	
XS750 E (UK and US) and XS750 SE (UK and US) –	
non EPA models...	10° BTDC at 1100 rpm
All other models...	10° BTDC at 1000 rpm
Advanced:	
XS750 E (US) and XS750 SE (US) –	
non EPA models	38.5° BTDC at 4000 rpm
XS750 E (UK), XS750 SE (UK) and	
XS750 F (US)	34° BTDC at 4000 rpm
XS750 E, SE and SF – EPA models	35° BTDC at 5000 rpm
Advance begins	2000 rpm
Pick-up coil resistance	560 ohms ± 20% at 20°C (68°F)

Spark plug

Make	NGK or Champion
Type	BP - 7ES or N - 7Y
Suppressor cap resistance	5 k ohms ± 15% at 20°C (68°F)

Ignition coil

Make	Hitachi
Type	CM11 – 54
Spark gap	6 mm at 1000 rpm
Primary resistance	2.75 ohms ± 10% at 20°C (68°F)
Secondary resistance	7.9 k ohms ± 20% at 20°C (68°F)

Specifications relating to Chapter 4

Front forks

Oil capacity per leg:	
XS750 E (UK)	200 cc (7.04/6.76 Imp/US fl oz)
XS750 E (US) and XS750 F (US)	190 cc (6.69/6.42 Imp/US fl oz)
All other models	234 cc (8.23/7.91 Imp/US fl oz)

Spring free length:
    XS750 E (UK and US) ... ...   503.2 mm (19.81 in)
    XS750 F (US) ... ...   498.0 mm (19.60 in)
    All other models ... ...   606.8 mm (23.89 in)

Rear suspension units
    Spring free length ... ...   256 mm (10.08 in)

Specifications relating to Chapter 5

Tyre pressures (cold)	Front	Rear
XS750 E and F:		
Solo	26 psi (1.8 kg/cm^2)	28 psi (2.0 kg/cm^2)
Pillion, or high speed riding	28 psi (2.0 kg/cm^2)	32 psi (2.3 kg/cm^2)
Pillion and extra load	40 psi (2.8 kg/cm^2)	40 psi (2.8 kg/cm^2)
XS750 E (EPA), SE (UK) and SF (US):		
Up to 198 lb (90 kg) load	26 psi (1.8 kg/cm^2)	28 psi (2.0 kg/cm^2)
198 − 410 lb (90 − 186 kg) load	28 psi (2.0 kg/cm^2)	32 psi (2.3 kg/cm^2)
High speed riding	40 psi (2.8 kg/cm^2)	40 psi (2.8 kg/cm^2)
XS750 SE (US):		
Up to 198 lb (90 kg) load	26 psi (1.8 kg/cm^2)	28 psi (2.0 kg/cm^2)
198 − 410 lb (90 − 186 kg) load	28 psi (2.0 kg/cm^2)	32 psi (2.3 kg/cm^2)
High speed riding	28 psi (2.0 kg/cm^2)	32 psi (2.3 kg/cm^2)

Note: *These pressures apply only when the original Bridgestone tyres are used. If tyres of a different type or size are fitted the values may differ.*

Brakes (front and rear)
Pad thickness:
    XS750 E and F (US) ... ...   6.5   mm (0.26 in)
    Wear limit ... ...   1.5   mm (0.006 in)
    XS750 E (UK) ... ...   11.0   mm (0.43 in)
    Wear limit ... ...   6.0   mm (0.20 in)
    XS750 SE (UK) ... ...   —
    Wear limit ... ...   6.5   mm (0.26 in)
    XS750 SE and SF (US):
    Front ... ...   10.2   mm (0.40 in)
    Wear limit ... ...   5.7   mm (0.22 in)
    Rear ... ...   11.0   mm (0.43 in)
    Wear limit ... ...   6.0   mm (0.24 in)
Disc thickness ... ...   7.0   mm (0.28 in)
    Service limit ... ...   6.5   mm (0.26 in)
Disc maximum warpage ... ...   0.15 mm (0.006 in)

Specifications relating to Chapter 6

Starter motor
    Field coil resistance ... ...   0.01 ohm at 20°C (68°F)
    Armature coil resistance ... ...   0.007 ohm at 20°C (68°F)

Voltage control regulator
    Make ... ...   National or Shindengen (some XS750 F models)
    Type ... ...   RD 1143 or SH 233 − integrated circuit
    Regulated voltage ... ...   14.5 ± 0.3 volts

Rectifier
    Make ... ...   National or Shindengen (some XS750 F models)
    Type ... ...   RD 1143 or SH 233 − silicon full − wave

Alternator
    Field coil resistance:
    XS750 SE (UK) ... ...   3.5 ohms ± 10% at 20°C (68°F)

Bulbs
    Headlamp:
    XS750 SE (UK) and XS750 F (US) ... ...   60/55 Watt
    Tail/stop lamp:
    XS750 SE (UK) ... ...   5/21 Watt
    Flashing indicators:
    XS750 SE (UK) ... ...   21 Watt

Model dimensions

	XS750 E (UK)	XS750 E (US)	XS750 SE (UK)	XS750 SE, and SF (US)	XS750 F (US)
Height	1130 mm (44.49 in) or 1150 mm (45.28 in)	1175 mm (46.3 in)	1245 mm (49.0 in)	1240 mm (48.8 in)	1175 mm (46.3 in)
Width	725 mm (28.54 in) or 895 mm (35.24 in)	900 mm (35.4 in)	925 mm (36.4 in)	870 mm (34.3 in)	900 mm (35.4 in)
Length	2160 mm (85.04 in)	2155 mm (84.8 in)	2155 mm (84.8 in)	2170 mm (85.4 in)	2140 mm (84.3 in)
Wheel base	1470 mm (57.87 in)	1465 mm (57.7 in)	1485 mm (58.5 in)	1500 mm (59.1 in)	1450 mm (57.1 in)
Seat height	810 mm (31.89 in)	820 mm (32.3 in)	815 mm (32.1 in)	815 mm (32.1 in)	820 mm (32.3 in)
Ground clearance (minimum)	140 mm (5.5 in)	140 mm (5.5 in)	155 mm (6.1 in)	160 mm (6.3 in)	140 mm (5.5 in)
Weight	232 kg (512 lb)	232 kg (512 lb)	230 kg (507 lb)	230 kg (507 lb)	237 kg (522 lb)

1 Introduction

This Chapter covers the XS750 models carrying the suffix E, SE, F and SF; the letter S denoting the model to be a Special (American style) version and the letters E and F denoting the model year. The E and SE models were introduced into the USA in September of 1977 and superseded the following September by the F and SF models. The E model was not introduced into the UK until March of 1978 and was discontinued in May of 1980, this being the introduction date for the SE model into the UK. The following text covers only the design changes made to these model types. Any service information not included in this Chapter can be assumed to be identical to that given in the main text of this manual.

Examples of the design changes made to the above listed models include the introduction of a fully transistorised ignition system, the incorporation of a two-position choke into the carburettor assembly and two slightly differing versions of a three-way preloaded adjustable fork design, one for Standard and one for Special models. Special models also have a different design of front brake caliper unit to that fitted to all other models in the range. The F model is supplied with tubeless tyre/ wheel assemblies.

Minor modifications throughout the model range include changes to the engine, clutch, gearbox, fuel system and electrical system, all of which are covered in the following Sections.

2 Positioning the piston rings

1   After the rings have been fitted to each piston, they should be positioned as shown in the figure accompanying this text before the piston is inserted into the cylinder bore.

3 Clutch assembly: modifications

1   When dismantling or reassembling the clutch refer carefully to Fig. 1.8 and be sure to note which components are fitted to your machine, also note exactly the way in which each is fitted. Several modifications have been made which mean that apparently similar components may not be interchangeable between different models; seek the advice of a good Yamaha Service Agent if in doubt.
2   Note particularly that the large circlip (item 2) is no longer fitted, the primary driven sprocket now being located by a raised shoulder on the shock absorber. On all models from 1979

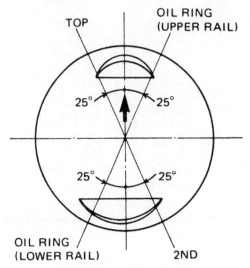

Fig. 7.1. Fitting the piston rings

onwards one or more adjustment shims may be fitted between the outer drum (item 10) and its outboard locating thrust washer (item 22) to control outer drum endfloat; refer to Section 4 of Chapter 8 of this Manual.

4 Gearbox: refitting the middle gear bearing

1   The middle gear journal ball bearing is now retained in position in the crankcase with four circlip halves instead of the previous two. Reference should be made to the figure accompanying this text for details of the modified design and it should be noted that it is not permissible to use the earlier design of bearing with one groove in the crankcase design with two grooves.

5 Camshaft chain: adjustment

1   When adjusting the camshaft chain, it should be noted that the adjuster bolt has been relocated from that position shown in the main text of the manual and is now on the right-hand (inboard) side of the tensioner boss. The method of adjustment remains identical to that given in Chapter 1.

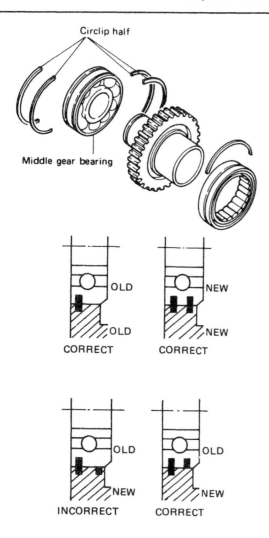

Circlip half

Middle gear bearing

OLD / OLD
CORRECT

NEW / NEW
CORRECT

OLD / NEW
INCORRECT

OLD / NEW
CORRECT

Fig. 7.2. Fitting the middle gear ball journal bearing

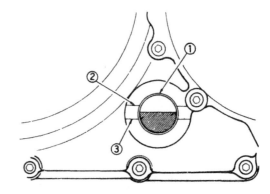

Fig. 7.3. Checking the engine/gearbox oil level

1   Sight glass   3   Minimum level mark
2   Maximum level mark

7.1 Add oil to the engine/gearbox unit through the crankcase filler point

6   Middle gear case: lubrication

1   The oil capacity of the middle gear case assembly fitted to the XS750 models covered in this Chapter differs from that of the original models. The revised capacity being 375 cc (13.20/ 12.68 Imp/US fl oz).

7   Engine/gearbox unit: checking the oil level

1   With the machine placed on its centre stand, start the engine and run it until normal operating temperature is reached. Stop the engine and allow the oil to settle before checking its level through the sight glass situated on the right-hand side of the engine, just to the rear of the lower part of the alternator cover. The oil level should be between the maximum and minimum level marks cast either side of the sight glass. If the oil level is below the minimum level mark it must be raised to the correct level before the engine is again started. Oil may be added by unscrewing and removing the filler plug located to the rear of the right-hand cylinder barrel and pouring oil through the filler point whilst keeping a close watch on the level in the sight glass. Do not forget to refit and tighten the filler plug on completion of adding the oil and wipe any spilt oil off the engine and cycle components.

8   Petrol taps: cleaning — XS750 SE, F and SF models

1   To clean the vacuum controlled petrol taps fitted to the above listed models, it is first necessary to lift the seat and remove the bolt securing the rear of the fuel tank. With the tap lever turned to the ON or RES position, lift the rear of the tank and detach the fuel line from the inboard facing stub of the tap body.
2   Remove the drain bolt and gasket situated at the base of the tap body and clean the bolt thoroughly in solvent to remove all traces of sediment. Check the condition of the gasket and renew it if it is damaged. Refit the bolt and gasket, reconnect the fuel line and relocate the tank and seat. Start the engine and check all disturbed connections on the tap body for signs of fuel leakage. Any leaks found must be cured before the machine is ridden.

9   Carburettors: 2-position choke

1   The XS750 models covered in this Chapter have basically the same type of carburettors as those shown in the main text, the only difference being the new two-position choke. This new choke moves from the vertical 'off' position downwards. At its furthest point of travel it is at the 'full on' position, from here

it is then moved back upward towards the vertical 'half on' position. The old type of choke had only a 'full on' position and an 'off' position and the choke lever moved from side-to-side rather than up and down.

2   If it becomes necessary to dismantle the carburettors, the procedure is very similar to that described in Chapter 2 which should be used, keeping in mind the slight operating difference and referring to the line drawing accompanying this text.

10 Carburettors: dismantling, examination, cleaning and reassembly

1   Remove the carburettors from the machine by using the procedure described in Chapter 2, Section 5 of this manual.
2   To separate the carburettors, remove the three screws which hold the choke shaft in position. Remove the choke shaft taking care not to dislodge the positioning balls on either side.
3   Remove the upper and lower brackets, noting the position of the synchronizing screws (this will be of aid when reassembling the carburettors). The carburettors may then be separated.
4   Dismantle each carburettor separately so that their parts are not accidentally interchanged. Remove the vacuum chamber cover, then the spring, needle fitting plate, jet needle, and diaphragm. When these components have been removed it will be seen that the diaphragm has a tab for which there is a matching recess in the carburettor body. This tab must be replaced into the recess when reassembling.
5   The choke assembly must then be removed from the left-hand side of the carburettor so that the starter jet can be removed, examined, and cleaned.
6   Remove the four screws which retain the float bowl cover and remove the float bowl. The main jet is located underneath this cover.

7   Remove the float pivot pin and the float assembly taking care not to lose the float valve needle which is located beneath the float level adjustment tag. Then remove the needle jet.
8   Inspect all parts for signs of wear and replace any damaged parts. Clean thoroughly in clean petrol and blow through the internal passageways to clear them. Do not use pointed objects or pieces of wire to clean any blockages as this will cause considerable damage.
9   To reassemble the carburettors, reverse the dismantling procedure, being careful to reinstall the diaphragm with its tab in the recess provided.

11 Air cleaner: removing, cleaning and refitting the element

1   The method of removing the air cleaner element fitted to the XS750 models covered in this Chapter differs from that given in the main text, although the actual construction and method of cleaning the element is similar.
2   To remove the element, first unscrew and remove the screw securing each one of the two side covers (SE and SF models only) to the air cleaner casing and remove each side cover from the machine. Loosen the wing bolt securing the element cover to the air cleaner casing and remove the cover complete with element by pulling it forwards and sideways to clear the casing.
3   After having cleaned or renewed the element (see Section 11, Chapter 2), refit the element together with its cover by using a reversal of the procedure given for removal. Check whilst locating the element that its seal is in good condition and seats correctly against the casing.
4   The revised period for cleaning the element is once every year or every 5000 miles (8000 km). This period should be reduced if the machine is ridden in constantly humid or dusty conditions.

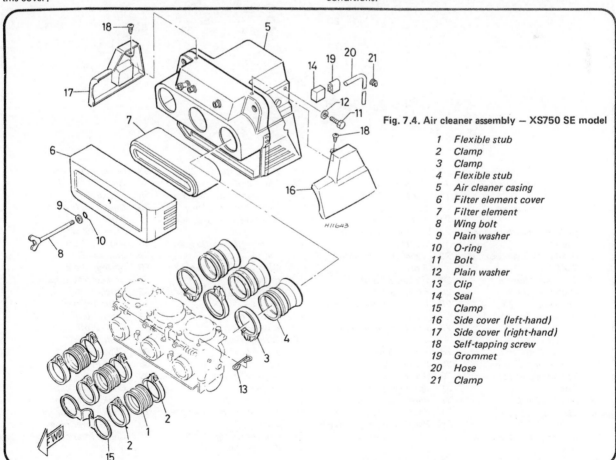

Fig. 7.4. Air cleaner assembly — XS750 SE model

1	Flexible stub
2	Clamp
3	Clamp
4	Flexible stub
5	Air cleaner casing
6	Filter element cover
7	Filter element
8	Wing bolt
9	Plain washer
10	O-ring
11	Bolt
12	Plain washer
13	Clip
14	Seal
15	Clamp
16	Side cover (left-hand)
17	Side cover (right-hand)
18	Self-tapping screw
19	Grommet
20	Hose
21	Clamp

Fig. 7.5. Carburettor assembly

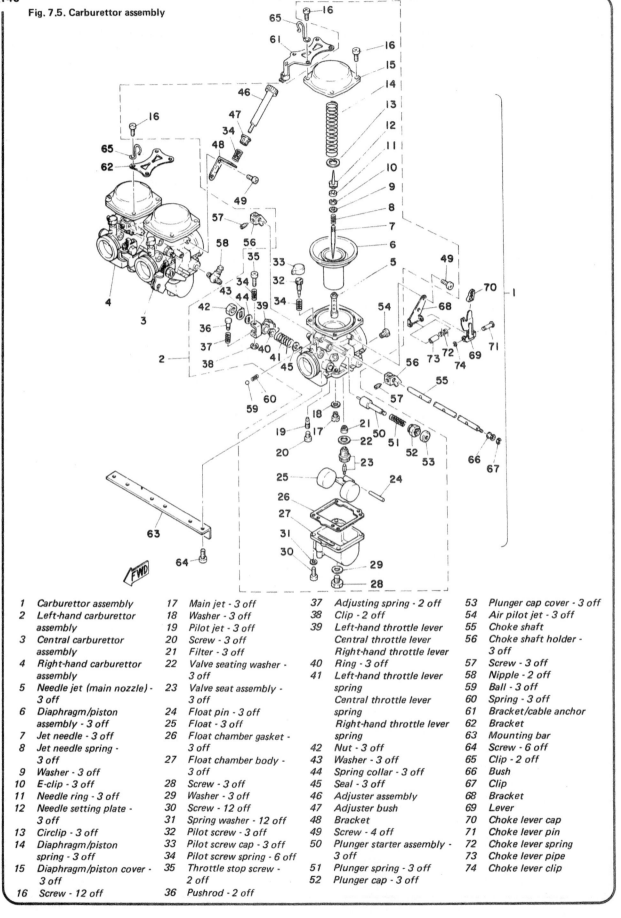

1	Carburettor assembly	17	Main jet - 3 off	37	Adjusting spring - 2 off	53	Plunger cap cover - 3 off
2	Left-hand carburettor assembly	18	Washer - 3 off	38	Clip - 2 off	54	Air pilot jet - 3 off
		19	Pilot jet - 3 off	39	Left-hand throttle lever	55	Choke shaft
3	Central carburettor assembly	20	Screw - 3 off		Central throttle lever	56	Choke shaft holder - 3 off
		21	Filter - 3 off		Right-hand throttle lever		
4	Right-hand carburettor assembly	22	Valve seating washer - 3 off	40	Ring - 3 off	57	Screw - 3 off
				41	Left-hand throttle lever spring	58	Nipple - 2 off
5	Needle jet (main nozzle) - 3 off	23	Valve seat assembly - 3 off			59	Ball - 3 off
		24	Float pin - 3 off		Central throttle lever spring	60	Spring - 3 off
6	Diaphragm/piston assembly - 3 off	25	Float - 3 off		Right-hand throttle lever spring	61	Bracket/cable anchor
		26	Float chamber gasket - 3 off			62	Bracket
7	Jet needle - 3 off	27	Float chamber body - 3 off	42	Nut - 3 off	63	Mounting bar
8	Jet needle spring - 3 off	28	Screw - 3 off	43	Washer - 3 off	64	Screw - 6 off
9	Washer - 3 off	29	Washer - 3 off	44	Spring collar - 3 off	65	Clip - 2 off
10	E-clip - 3 off	30	Screw - 12 off	45	Seal - 3 off	66	Bush
11	Needle ring - 3 off	31	Spring washer - 12 off	46	Adjuster assembly	67	Clip
12	Needle setting plate - 3 off	32	Pilot screw - 3 off	47	Adjuster bush	68	Bracket
13	Circlip - 3 off	33	Pilot screw cap - 3 off	48	Bracket	69	Lever
14	Diaphragm/piston spring - 3 off	34	Pilot screw spring - 6 off	49	Screw - 4 off	70	Choke lever cap
		35	Throttle stop screw - 2 off	50	Plunger starter assembly - 3 off	71	Choke lever pin
15	Diaphragm/piston cover - 3 off			51	Plunger spring - 3 off	72	Choke lever spring
		36	Pushrod - 2 off	52	Plunger cap - 3 off	73	Choke lever pipe
16	Screw - 12 off					74	Choke lever clip

11.2 Remove the air filter cover complete with element

12 Electronic ignition system: method of operation

1   The electronic ignition system used on the XS750 models covered in this Chapter combines an ignition pulser coil and a transistor switching circuit in order to break the flow of current to the primary ignition coil. This type of system means that mechanical contact breakers are no longer needed to induce a high-surge voltage in the secondary circuit by breaking the primary circuit. Electronic ignition helps to ensure better ignition at all times for a much longer time. As no mechanical contact breaker is used, correct ignition timing is maintained for far longer as the electrical system does not wear to any extent. This system is also much easier to maintain. A stronger spark is produced by the electronic system and this also helps to give a better performance.

2   The electronic ignition system consists of a pick-up unit, and an ignitor unit. The former is made up of a pick-up coil rotor, and a governor, the governor being the conventional mechanical type. The ignitor unit consists of a transistor switching circuit which interrupts the flow of current in the primary circuit of the ignition coil.

3   The system operates by means of a projection on the rotor attached to the crankshaft. As the projection passes the pick-up coil core a pulse is generated in the pick-up coil and a current flows to the transistor switching circuit and then to the distributing circuit. To prevent the flow of primary current to the ignition coil if the engine should stall, the pulser circuit continues to function while the machine is at rest, effectively 'turning off' the circuit even though the main ignition switch is still turned on.

13 Electronic ignition system: fault diagnosis

1   Should the electronic ignition system fail to operate, check the system by employing the following procedure. A multimeter set to the resistance function will be needed for these tests.

2   Test the system for faulty connections or a break in the wiring circuit. Test the ignition coils both for resistance and continuity in the primary and secondary windings. Compare the readings obtained on the meter with those given in the Specifications at the beginning of this Chapter.

3   Finally, test the pick-up coil windings both for continuity and resistance and compare the reading obtained with that given in Specifications.

4   If the above tests prove to be satisfactory, then it may be assumed that the ignitor unit is defective in which case the unit should be replaced with a serviceable item. If in doubt as to the condition of the system or how to carry out any of the above listed tests, it is recommended that the advice of a Yamaha repair specialist or a qualified auto-electrician be sought.

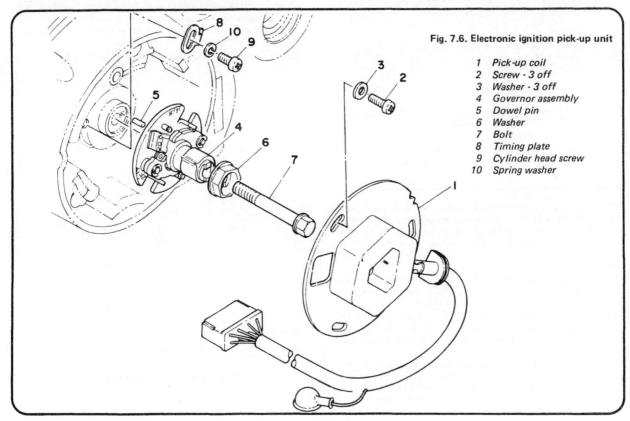

Fig. 7.6. Electronic ignition pick-up unit

1   *Pick-up coil*
2   *Screw - 3 off*
3   *Washer - 3 off*
4   *Governor assembly*
5   *Dowel pin*
6   *Washer*
7   *Bolt*
8   *Timing plate*
9   *Cylinder head screw*
10 *Spring washer*

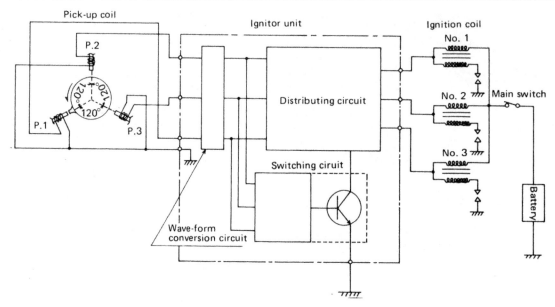

Fig. 7.7. Electronic ignition system — circuit diagram

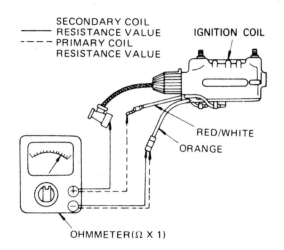

Fig. 7.8. Testing the ignition coil primary and secondary windings for resistance

14 Electronic ignition system: checking and adjusting the ignition timing

1   The procedure for checking and adjusting the ignition timing on the electronic ignition system is a relatively quick and simple operation which requires the use of a stroboscopic timing light. Reference should be made to the Specifications at the beginning of this Chapter for information on the ignition timing for each model type.

2   Commence the timing procedure by connecting the strobe to the left-hand (No 1) cylinder HT lead. Start the engine and set it to run at the specified engine speed (1000 or 1100 rpm). Direct the strobe beam towards the '1F' mark on the ATU. This mark should align with the stationary pointer if the timing is correct. If the pointer and marks do not align, the timing is in need of adjustment.

3   To adjust the timing, stop the engine and loosen the three pick-up base plate retaining screws. Restart the engine and move the unit until the marks align in the strobe beam. Stop the engine and re-tighten the screws. Do **NOT** bend the adjusting pointer.

4   After tightening the screws, recheck the timing again for No. 1 cylinder.

5   Raise the engine speed to that specified (4000 or 5000 rpm). The pointer should begin to move towards the 'full advance' marks on the governor as the engine speed reaches 2000 rpm and be in alignment with the area indicated by the 'full advance' marks as the specified engine speed is reached.

6   If a stroboscopic timing light is not readily available, the timing may be checked conveniently, if less accurately, as a static setting by using a dial gauge. Refer to the Specifications for the degree figures for each particular model.

7   Information on the removal and refitting of the pick-up coil and governor assemblies and the setting of the pick-up base plate may be found in Section 13 of Chapter 8.

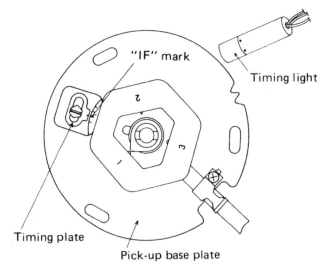

Fig. 7.9. Ignition timing adjustment

14.2 Note the position of the '1F' mark with the pointer

14.3 Loosen the pick-up base plate retaining screws to adjust the timing

15 Front forks: adjustment

1   The XS750 models covered in this Chapter are fitted with front forks which incorporate a method of adjusting the tension of each fork spring. The tension of each fork spring can be altered by removing the rubber cap at the top of the fork leg and then using a large screwdriver placed in the spring adjuster to depress the adjuster and turn it to obtain the desired spring tension. There are three possible positions but it should be remembered that both fork spring adjusters must be set in the same position.

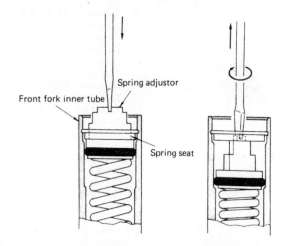

Front fork inner tube

Spring adjustor

Spring seat

Fig. 7.10. Front fork spring pre-load adjustment

16 Front forks: changing the fork oil

1   Before draining any oil from each front fork leg, it is necessary to position the machine so that the fork legs are fully extended. This may be achieved by placing the machine on its main stand and positioning a block of wood beneath the front of the engine crankcase to raise the front wheel clear of the ground. Alternatively, the front wheel may be removed from the machine.
2   Remove the rubber cap from the top of each fork leg and service each fork leg by using the following procedure.

3   The spring adjuster, spring seat, and fork spring are retained by a spring clip. To remove this spring clip it is first necessary to depress the components beneath it. Having done this it may be removed with a small screwdriver or a pair of small nosed pliers. Remember that the components beneath the clip are compressed and will spring out once the pressure upon them has been released unless the clip is removed gently. Withdraw the components from the stanchion.
4   Place a container under the drain plug hole in the fork leg and then remove the drain screw. When nearly all the oil has drained out, pump out the remaining oil by slowly raising and lowering the fork leg.
5   On completion of draining the fork leg of oil, inspect the drain screw gasket for damage and renew it if necessary before refitting and tightening the drain screw. Note that any oil drained from the fork leg must be kept clear of the brake disc. Should the disc become contaminated, it must be wiped clean with a petrol-moistened rag and allowed to dry before the machine is ridden.
6   Refer to the Specifications at the beginning of this Chapter and pour the correct amount of the recommended fork oil into the fork stanchion. Having refilled the stanchion, slowly pump the lower fork leg up and down to distribute the oil evenly.
7   Inspect the O-ring on the spring adjuster and renew it if it is in any way damaged. Refit the fork spring in the stanchion, then the O-ring, spring adjuster, spring seat, a new spring clip, and finally the rubber cover.

17 Front wheel: removal and refitting — XS750 SE and SF models

1   Before removing the front wheel from either of the above listed models, it is first necessary to remove the left-hand brake caliper from its fork leg mounting point by following the procedure listed in paragraph 3 of the following Section.
2   With the caliper unit tied to a point clear of the wheel, remove the split-pin from the wheel spindle nut and remove the nut. Remove the speedometer drive cable securing bolt from the wheel drive box and loosen the pinch bolt at the base of the right-hand fork leg in order to free the spindle. With the machine properly supported, withdraw the wheel spindle and move the wheel clear of the fork legs.
3   It is worth noting that directly the brake disc is moved clear of the pads, there is a danger that inadvertent operation of the brake lever will cause the pistons to be forced out of the caliper bores. To prevent this happening, a wooden wedge should be inserted between the brake pads directly the wheel is removed.

4 Refitting of the wheel is essentially a reversal of the removal procedure, noting the following points. Take great care when refitting the wheel to ensure that the brake pads are far enough apart to allow the brake disc to pass between them. Failure to observe this will result in the pads becoming damaged by the disc.

5 Ensure that the projecting stub on the inner face of the fork leg is located correctly between the two projections on the speedometer drive box before inserting the wheel spindle. Tighten the spindle securing nut to a torque of 10.5 kg m (76 lb ft) and fit a new split-pin.

6 Check the front forks for correct operation by pumping them up and down several times. Ensure that the clearance between the caliper support bracket of the right-hand fork leg and the sides of the brake disc is of an equal amount either side of the disc. This clearance may be determined by moving the fork leg back and forth. Once these clearances are equal, tighten the pinch bolt to a torque of 2.0 kg m (14.5 lb ft).

7 On completion of refitting the wheel, check and tighten both caliper to fork leg securing bolts to a torque of 2.5 kg m (18 lb ft) and check the brakes for correct operation and signs of fluid leakage before riding the machine.

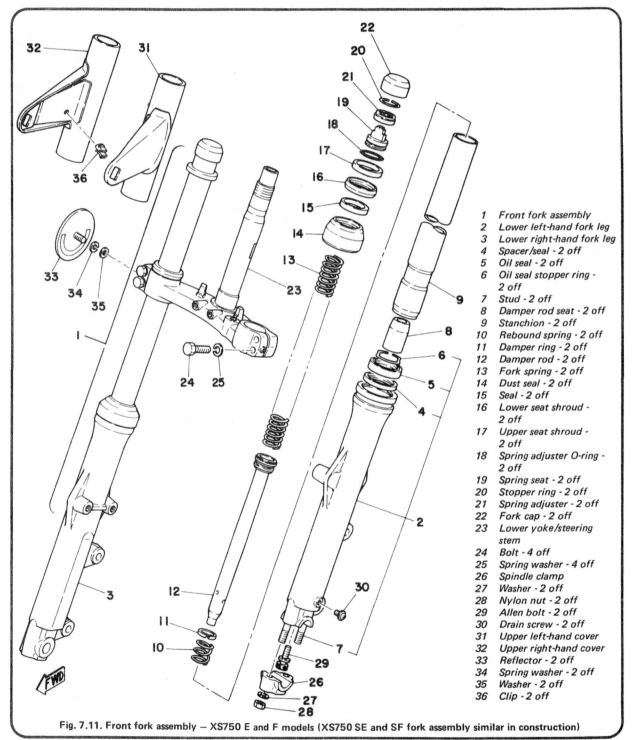

1   Front fork assembly
2   Lower left-hand fork leg
3   Lower right-hand fork leg
4   Spacer/seal - 2 off
5   Oil seal - 2 off
6   Oil seal stopper ring - 2 off
7   Stud - 2 off
8   Damper rod seat - 2 off
9   Stanchion - 2 off
10 Rebound spring - 2 off
11 Damper ring - 2 off
12 Damper rod - 2 off
13 Fork spring - 2 off
14 Dust seal - 2 off
15 Seal - 2 off
16 Lower seat shroud - 2 off
17 Upper seat shroud - 2 off
18 Spring adjuster O-ring - 2 off
19 Spring seat - 2 off
20 Stopper ring - 2 off
21 Spring adjuster - 2 off
22 Fork cap - 2 off
23 Lower yoke/steering stem
24 Bolt - 4 off
25 Spring washer - 4 off
26 Spindle clamp
27 Washer - 2 off
28 Nylon nut - 2 off
29 Allen bolt - 2 off
30 Drain screw - 2 off
31 Upper left-hand cover
32 Upper right-hand cover
33 Reflector - 2 off
34 Spring washer - 2 off
35 Washer - 2 off
36 Clip - 2 off

Fig. 7.11. Front fork assembly — XS750 E and F models (XS750 SE and SF fork assembly similar in construction)

18 Front disc brake: checking and renewing the pads — XS750 SE and SF models

1   The front brake caliper assembly fitted to the above listed models differs from that fitted to the other XS750 models inasmuch as it is no longer necessary to dismantle the caliper unit mounting bracket assembly in order to gain access to the pads as described in paragraph 2, Section 3, Chapter 5.
2   The amount of wear on the brake pads may be checked by viewing the caliper assembly from the rear. A wear indicator is attached to each pad.
3   To renew the brake pads, remove the plastic cap which covers the head of the caliper to fork leg securing bolt. Support the caliper and unscrew and withdraw the bolt, allowing the caliper to drop clear of the fork leg. With the caliper cupped in one hand so that there is no strain imposed on the hose connections, remove the coil spring followed by the pad securing pin. A pair of long-nose pliers is the best tool for removing these items.
4   Withdraw the brake pads from the caliper and replace them with a new set. Note that as well as renewing the pads, the coil spring and pin must also be renewed. Refitting the caliper assembly to the fork leg is a reversal of the removal procedure.
5   It will be found that a new brake caliper assembly will have a spring fitted in order to retain the pads in position. This spring must be discarded before the assembly is fitted to the machine.

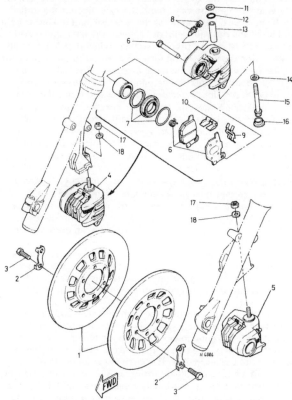

Fig. 7.12. Front brake caliper assembly — XS750 SE and SF models

1	Brake disc	9	Pad spring
2	Lock washer	10	Retainer plate
3	Bolt	11	Plate washer
4	Caliper assembly (right-hand)	12	O-ring
		13	Sleeve
5	Caliper assembly (left-hand)	14	Plate washer
		15	Bolt
6	Pad set	16	Cap
7	Seal kit	17	Nut
8	Bleed screw and cap	18	Spring washer

19 Front brake: adjusting the brake lever free play

1   It is recommended that the amount of free play, measured at the end of the front brake lever, is checked at 6 monthly intervals or after every 2500 miles (4000 km). The amount of free play should be 5 — 8 mm (0.2 — 0.3 in). If found to be incorrect, adjustment may be made by pulling back the rubber gaiter from the brake lever pivot point so as to expose the adjuster screw and locknut threaded into the stub of the lever. Loosen the locknut and turn the screw in or out until the correct amount of free play is obtained before the end of the screw comes into contact with the piston of the master cylinder. Hold the screw in position and tighten the locknut. Recheck the amount of free play and if correct, refit the gaiter.

19.1 Pull back the rubber gaiter to expose the brake lever adjuster screw

20 Rear wheel: removal and refitting

1   The procedure for removing and refitting the rear wheel assembly fitted to the XS750 models covered in this Chapter is similar to that given in paragraphs 3 and 4, Section 9, Chapter 4; the following points should, however, be noted.
2   Less effort will be required to compress the rear suspension in order to fit the wire stay if the left-hand rear suspension unit is removed beforehand.
3   Never depress the rear brake pedal whilst the wheel is removed. Should the pedal be depressed, the piston will be forced out of the caliper bore. To prevent this happening, a wooden wedge should be inserted between the brake pads directly the wheel is removed.
4   Take great care when refitting the wheel to ensure that the brake pads are far enough apart to allow the brake disc to pass between them. Failure to observe this will result in the pads becoming damaged. Always fit a new split-pin through the wheel spindle nut after having torque loaded the nut to 15.0 kg m (108 lb ft) and the pinch bolt to 0.60 kg m (4.0 lb ft).

21 Rear brake: adjusting the brake pedal free play

1   It should be noted that unless the correct amount of brake pedal free play is maintained, there is a possibility of excessive brake drag occurring. It is recommended that the amount of free play, measured at the end of the brake pedal as shown in the accompanying figure, is checked at 6 monthly intervals or after every 2500 miles (4000 km). The amount of free play should be 13 — 15 mm (0.51 — 0.59 in).

2   Before adjusting the amount of pedal free play, check that the pedal is at its correct height in relation to the top edge of the footrest rubber. The measurement of pedal height is shown in the accompanying figure and should be 17 − 23 mm (0.67 − 0.91 in). To adjust the position of the brake pedal, loosen the adjuster bolt locknut and turn the adjuster bolt clockwise or anti-clockwise to raise or lower the pedal. With the pedal at its correct height, tighten the locknut and adjust the pedal for free play as follows.

3   Loosen the brake rod adjuster locknut and screw the brake rod downwards until there is noticeable free play between the rod and the master cylinder. Turn the brake rod in until it comes into contact with the master cylinder and then turn it out approximately 1 1/5 turns (SE and SF models) or 1 3/4 turns (E and F models) to obtain the correct amount of free play. Retighten the locknut and check that the punch mark on the brake rod does not appear above the top of the locknut.

4   Remember that adjustment of the brake pedal may necessitate readjustment of the rear stop lamp switch. Refer to Section 18 of Chapter 6 for details of switch adjustment.

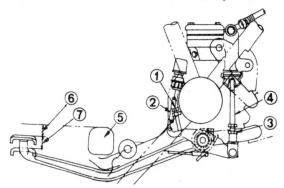

Fig. 7.13. Adjusting the rear brake pedal free play

1	Adjuster bolt (pedal height)	4	Brake rod
2	Locknut	5	Footrest
3	Locknut	6	Brake pedal height
		7	Brake pedal free play

22 Tyre pressures

1   The maintenance of correct tyre pressures is important on all motorcycles to ensure the best possible handling and road-holding at all times. Incorrectly inflated tyres may cause an accident through loss of control, either as a result of a moment of over-enthusiasm, or in an emergency. On large touring motor-cycles the speeds at which the machine is used and the loads it carries may differ widely from journey to journey, and in many cases one set of tyre pressures will not be suitable for all applications. For this reason the manufacturers recommend variations as shown in the table in the Specifications at the beginning of this Chapter. All pressures relate to cold tyres.

23 Tubeless tyres: general information, removal and refitting — XS750 F model

1   The XS750 F model is fitted with tubeless tyres as standard equipment. These tyres are marked 'TUBELESS' on the tyre sidewall, whereas the cast aluminium wheels are marked 'SUITABLE FOR TUBELESS TYRES' on one of the wheel spokes. On no account should tubeless tyres be fitted to a wheel designed to take only tyres that are fitted with an inner tube. If this is attempted, the tyre will most certainly suddenly deflate during use causing failure of the tyre with the possibility of serious personal injury.

2   The primary advantage of fitting tubeless tyres is that they will accept penetration by sharp objects, such as nails etc, without loss of air. Even if loss of air is experienced, because there is no inner tube to rupture, in normal conditions a sudden blow-out is avoided.

3   It is strongly recommended that should a repair to a tubeless tyre be necessary, the wheel is removed from the machine and taken to a tyre fitting specialist who is willing to do the job or alternatively, taken to an official Yamaha dealer. This is because the force required to break the seal between the wheel rim and tyre bead is considerable and considered to be beyond the capabilities of an individual working with normal tyre removing tools. Any abortive attempt to break the rim to bead seal may also cause damage to the wheel rim, resulting in an expensive wheel replacement. If, however, a suitable bead releasing tool is available, and experience has already been gained in its use, tyre removal and refitting can be accomplished as follows.

4   With the wheel removed from the machine, deflate the tyre by removing the valve insert and when it is fully deflated, push the bead of the tyre away from the wheel rim on both sides so that the bead enters the centre well of the rim. As noted, this operation will almost certainly require the use of a bead releasing tool.

5   Insert a tyre lever close to the valve and lever the edge of the tyre over the outside of the wheel rim. Very little force should be necessary; if resistance is encountered it is probably due to the fact that the tyre beads have not entered the well of the wheel rim all the way round the tyre. Should the initial problem persist, lubrication of the tyre bead and the inside edge and lip of the rim will facilitate removal. Use a recommended lubricant, a dilute solution of washing-up liquid or french chalk. Lubrication is usually recommended as an aid to tyre fitting but its use is equally desirable during removal. The risk of lever damage to wheel rims can be minimised by the use of pro-prietary plastic rim protectors placed over the rim flange at the point where the tyre levers are inserted. Suitable rim protectors may be fabricated very easily from short lengths (4 − 6 inches) of thick-walled nylon petrol pipe which have been split down one side using a sharp knife. The use of rim pro-tectors should be adopted whenever levers are used and, therefore, when the risk of damage is likely, Once the tyre has bead opposite the working area is always in the well. Apply wheel rim so that the tyre is completely free on one side.

6   Working from the other side of the wheel, ease the other edge of the tyre over the outside of the wheel rim which is furthest away. Continue to work around the rim until the tyre is freed completely from the rim.

7   Refitting of the tyre is virtually a reversal of the removal procedure. If the tyre has a balance mark (usually a spot of coloured paint), this must be positioned alongside the valve. Similarly, any arrow indicating direction of rotation must face the right way.

8   Starting at the point furthest from the valve, push the tyre bead over the edge of the wheel rim until it is located in the central well. Continue to work around the tyre in this fashion until the whole of one side of the tyre is on the rim. It may be necessary to use a tyre lever during the final stages. Here again, the use of a lubricant will aid fitting. It is recommended strongly that when refitting the tyre only a recommended lubricant is used because such lubricants also have sealing properties. Do not be over generous in the application of lubricant or tyre creep may occur.

9   Fitting the upper bead is similar to fitting the lower bead. Start by pushing the bead over the rim and into the well at a point diametrically opposite the tyre valve. Continue working round the tyre, each side of the starting point, ensuring that the been edged over the wheel rim, it is easy to work around the lubricant as necessary. Avoid using tyre levers unless absolutely essential, to help reduce damage to the soft wheel rim. The use of the levers should be required only when the final portion of bead is to be pushed over the rim.

10 Lubricate the tyre beads again prior to inflating the tyre,

and check that the wheel rim is evenly positioned in relation to the tyre beads. Inflation of the tyre may well prove impossible without the use of a high pressure air hose. The tyre will retain air completely only when the beads are firmly against the rim edges at all points and it may be found when using a foot pump that air escapes at the same rate as it is pumped in. This problem may also be encountered when using an air hose, on new tyres which have been compressed in storage and by virtue of their profile hold the beads away from the rim edges. To overcome this difficulty, a tourniquet may be placed around the circumference of the tyre, over the central area of the tread. The compression of the tread in this area will cause the beads to be pushed outwards in the desired direction. The type of tourniquet most widely used consists of a length of hose closed at both ends with a suitable clasp fitted to enable both ends to be connected. An ordinary tyre valve is fitted at one end of the tube so that after the hose has been secured around the tyre it may be inflated, giving a constricting effect. Another possible method of seating beads to obtain initial inflation is to press the tyre into the angle between a wall and the floor. With the airline attached to the valve additional pressure is then applied to the tyre by the hand and shin, as shown in the accompanying illustration. The application of pressure at four points around the tyre's circumference whilst simultaneously applying the airhose will often effect an initial seal between the tyre beads and wheel rim, thus allowing inflation to occur.

11 Having successfully accomplished inflation, increase the pressure to 40 psi and check that the tyre is evenly disposed on the wheel rim. This may be judged by checking that the thin positioning line found on each tyre wall is equidistant from the rim around the total circumference of the tyre. If this is not the case, deflate the tyre, apply additional lubrication and reinflate. Minor adjustments to the tyre position may be made by bouncing the wheel on the ground.

12 Always run the tyre at the recommended pressures and never under or over-inflate. The correct pressures for solo use, for high speed riding and for riding with the machine carrying a load, are listed in the preceding Section.

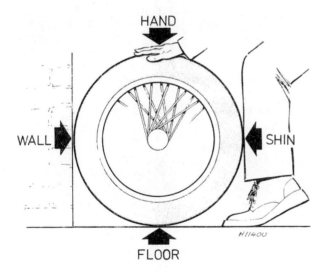

Fig. 7.14. Method of tubeless tyre inflation

Apply pressure at the four marked points to facilitate initial inflation

24 Tubeless tyres: repair — XS750 F model

1 Should a tubeless tyre become punctured, it must be removed for inspection for damage before any attempt is made at remedial action. The temporary repair of a punctured tyre by inserting a plug from the outside should not be attempted. Although this type of temporary repair is used widely on cars, the manufacturers strongly recommend that no such repair is carried out on a motorcycle tyre. Not only does the tyre have a thinner carcass, which does not give sufficient support to the plug, but the consequences of a sudden deflation are often sufficiently serious that the risk of such an occurrence should be avoided at all costs.

2 The tyre should be inspected both inside and out for damage to the carcass. Unfortunately the inner lining of the tyre — which takes the place of the inner tube — may easily obscure any damage and some experience is required in making a correct assessment of the tyre condition.

3 There are two main types of tyre repair which are considered safe for adoption in repairing tubeless motorcycle tyres. The first type of repair consists of inserting a mushroom-headed plug into the hole from the **inside** of the tyre. The hole is prepared for insertion of the plug by reaming and the application of an adhesive. The second repair is carried out by buffing the inner lining in the damaged area and applying a cold or vulcanised patch. Because both inspection and repair, if they are to be carried out safely, require experience in this type of work, it is recommended that the tyre be placed in the hands of a repairer with the necessary skills, rather than repaired in the home workshop.

4 In the event of an emergency, the only recommended 'get-you-home' repair is to fit a standard inner tube of the correct size. If this course of action is adopted, care should be taken to ensure that the cause of the puncture has been removed before the inner tube is fitted.

5 In the event of the unavailability of tubeless tyres, ordinary tubed tyres fitted with inner tubes of the correct size may be fitted. Refer to the manufacturer or a tyre fitting specialist to ensure that only a tyre and tube of equivalent type and suitability is fitted, and also to advise on the fitting of a valve nut to the rim hole.

25 Tubeless tyres: tyre valve and valve core renewal — XS750 F model

1 It will be appreciated from the preceding Sections that the adoption of tubeless tyres has made it necessary to modify the valve arrangement, as there is no longer an inner tube which can carry the valve core. The problem has been overcome by fitting a separate tyre valve which passes through a close-fitting hole in the rim, and which is secured by a nut and locknut. The valve is fitted from the rim well, and it follows that the valve can be removed and replaced only when the tyre has been removed from the rim. Leakage of air from around the valve body is likely to occur only if the sealing seat fails or if the nut and locknut become loose.

2 The valve core is of the same type as that used with tubed tyres, and screws into the valve body. The core can be removed with a small slotted tool which is normally incorporated in plunger type pressure gauges. Some valve dust caps incorporate a projection for removing valve cores. Although tubeless tyre valves seldom give trouble, it is possible for a leak to develop if a small particle of grit lodges on the sealing face. Occasionally, an elusive slow puncture can be traced to a leaking valve core, and this should be checked before a genuine puncture is suspected.

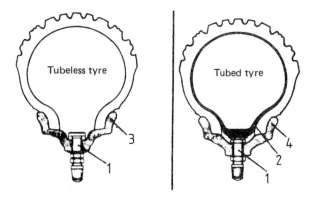

Fig. 7.15. Profile of tubeless and tubed type tyres

1   Tyre valve
2   Inner tube
3   Wheel (for tubeless tyre)
4   Wheel (for tubed tyre)

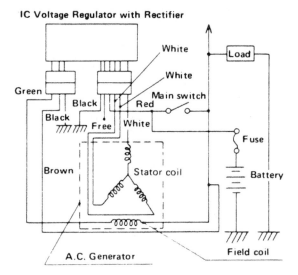

Fig. 7.16. Charging circuit diagram

26 Voltage regulator: testing

1   The voltage regulator fitted to the XS750 models covered in this Chapter is of the integrated circuit (IC) type and is contained within the same casing as the rectifier. This type of regulator is non-adjustable and because of its construction, much lighter than the mechanical type of regulator fitted to the earlier XS750 models.

2   Before testing the regulator, check the condition of the battery by using a hydrometer to measure the specific gravity of the electrolyte. If the reading obtained shows the specific gravity to be lower than 1.260, the battery must be recharged and the specific gravity rechecked to ensure it is between 1.260 and 1.280.

3   Carry out an inspection of the battery terminals and the regulator block connectors for both looseness and signs of corrosion. Check the wiring for signs of chafing against any cycle components which may lead to a short circuit. Once the circuit components have been thoroughly inspected for signs of damage or deterioration the regulator can be checked as follows.

4   Refer to the figure accompanying this text and connect two voltmeters to the block connector terminals as illustrated. Take great care not to short circuit the system when connecting the voltmeters.

5   Switch on the main ignition but not the lights or indicators. The reading shown on voltmeter V^2 should be less than 1.8 volts. Start the engine and allow it to idle. Gradually increase the engine speed whilst observing the readings on both voltmeters. The voltmeter V^2 should give a gradual increase in reading up to 9 – 11 volts as the engine speed increases. The voltmeter V^1 should maintain a reading of 14.2 – 14.8 volts throughout the engine speed range.

6   Should the above readings not be maintained, the regulator is defective and should be renewed.

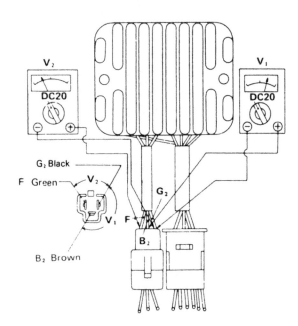

Fig. 7.17. Testing the voltage regulator

27 Silicon rectifier: testing

1   The silicon rectifier is contained within the same casing as the IC voltage regulator. To test the rectifier, set a multimeter to the resistance function and refer to the table and figure accompanying this text. Test each rectifier element by placing the positive (+) and negative (−) probes of the multimeter on the terminals of the two block connectors as indicated. If one of the elements is found to be unserviceable, the rectifier unit must be renewed.

2   It should be noted that it is possible for the rectifier to become damaged if it is subjected to overcharging. The rectifier should never be connected directly to the battery when carrying out a check for continuity. Always take care to avoid causing a

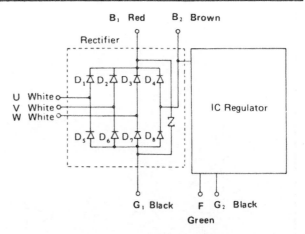

Fig. 7.18. Silicon rectifier circuit diagram

short circuit when placing the probes of the multimeter on the terminals of the block connectors and never inadvertently reverse the positive and negative leads at the battery terminals.

28 Reserve lighting system: general description and operation

1   This system has two functions, the first being to inform the rider that one of the headlamp filaments is inoperative and the second being to switch current from the inoperative filament to the remaining serviceable filament.

2   The system is located beneath the fuel tank and is connected into the headlamp circuit only. On the SE and SF models and the later E models, a safety relay is incorporated. This ensures that the headlamp is switched on automatically when the engine is started, even with the headlamp switch in the 'Off' position The F models incorporate a function whereby the headlamp comes on automatically when the engine is started and stays on until the main switch is turned to the 'Off' position, even if the engine has stalled.

3   It is recommended that any testing of the system be entrusted to an official Yamaha dealer who will have the necessary equipment to trace any faults within the system in a relatively short space of time and the facilities required to renew any units that may be found to be defective.

29 Headlamp: replacing the bulb – XS750 F model

1   The XS750 F model is the only US model to be equipped with a separate bulb in the headlamp lens assembly as opposed to the sealed beam unit fitted to earlier models. The method of bulb removal is similar to that described in Section 10 of Chapter 6. Reference should be made to the figure accompanying this text for details of the lens assembly.

2   Care should be taken whilst handling the bulb to avoid touching the glass envelope. The oil and moisture passed from the hand to the glass will shorten the life of the bulb and adversely affect the transparency of the glass. The glass may be cleaned with a soft cloth moistened with alcohol.

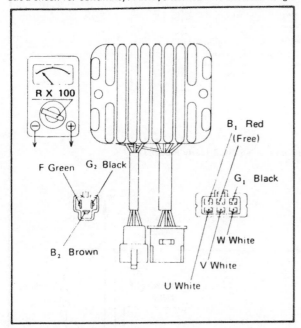

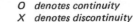

Fig. 7.19. Testing the silicon rectifier

O   denotes continuity
X   denotes discontinuity

Checking element	Pocket tester connecting point		Good	Replace (element shorted)	Replace (element opened)
	(+) Red	(−) Black			
D₁	B₁	U	O	O	×
	U	B₁	×	O	×
D₂	B₁	V	O	O	×
	V	B₁	×	O	×
D₃	B₁	W	O	O	×
	W	B₁	×	O	×
D₄	B₁	B₂	O	O	×
	B₂	B₁	×	O	×
D₅	U	G₁	O	O	×
	G₁	U	×	O	×
D₆	V	G₁	O	O	×
	G₁	V	×	O	×
D₇	W	G₁	O	O	×
	G₁	W	×	O	×
D₈	B₂	G₁	O	O	×
	G₁	B₂	×	O	×

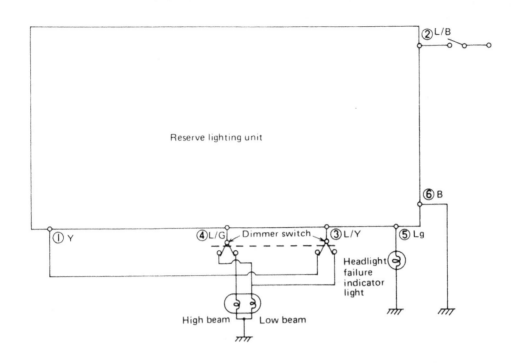

Fig. 7.20. Reserve lighting system circuit diagram — XS750 E (early models)

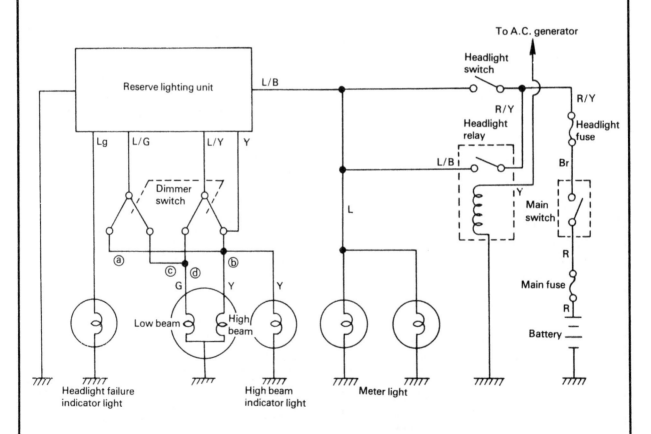

Fig. 7.21. Reserve lighting system circuit diagram — XS750 E (late models)

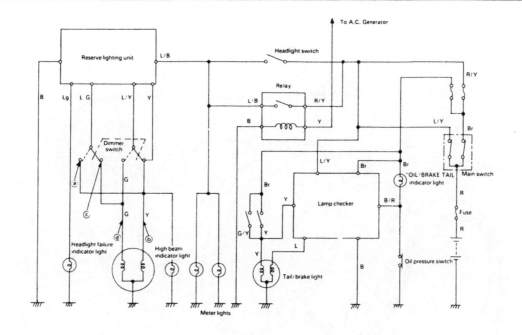

Fig. 7.22. Reserve lighting system circuit diagram — XS750 SE and SF models

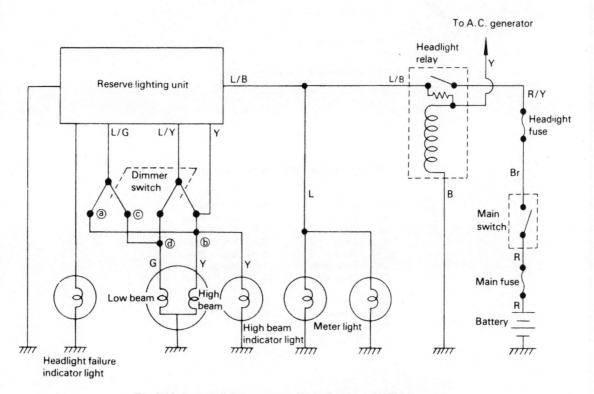

Fig. 7.23. Reserve lighting system circuit diagram — XS750 F model

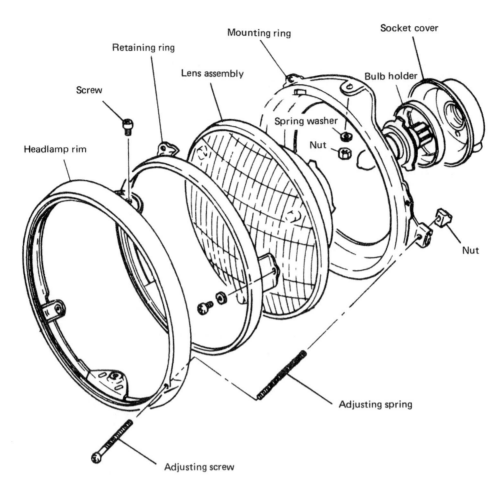

Fig. 7.24. Headlamp lens assembly — XS750 F model

Colour code for XS750 wiring diagrams

R	Red	L/W	Blue/White
Br	Brown	R/L	Red/Blue
W	White	L/Y	Blue/Yellow
Lg	Light green	L/G	Blue/Green
B	Black	Y/B	Yellow/Black
Y	Yellow	Br/W	Brown/White
Dg	Dark green	Y/G	Yellow/Green
Ch	Chocolate	W/G	White/Green
L	Blue	Y/R	Yellow/Red
Gy	Gray	W/R	White/Red
O	Orange	G/R	Green/Red
R/W	Red/White		

159

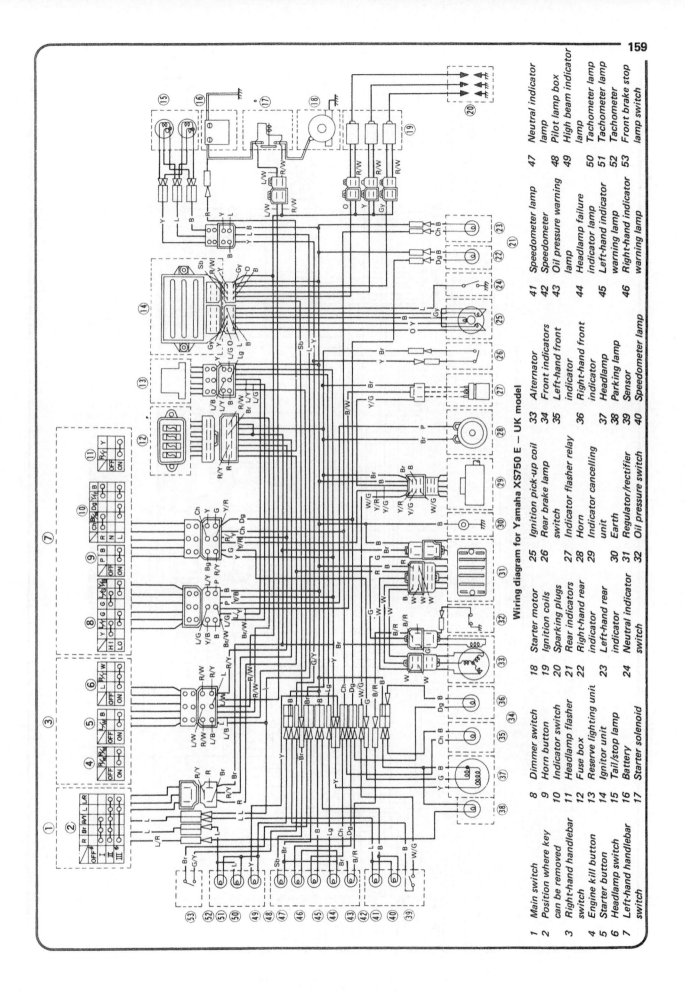

Wiring diagram for Yamaha XS750 E — UK model

1	Main switch
2	Position where key can be removed
3	Right-hand handlebar switch
4	Engine kill button
5	Starter button
6	Headlamp switch
7	Left-hand handlebar switch
8	Dimmer switch
9	Horn button
10	Indicator switch
11	Headlamp flasher
12	Fuse box
13	Reserve lighting unit
14	Ignitor unit
15	Tail/stop lamp
16	Battery
17	Starter solenoid
18	Starter motor
19	Ignition coils
20	Sparking plugs
21	Rear indicators
22	Right-hand rear indicator
23	Left-hand rear indicator
24	Neutral indicator switch
25	Ignition pick-up coil
26	Rear brake lamp switch
27	Indicator flasher relay
28	Horn
29	Indicator cancelling unit
30	Earth
31	Regulator/rectifier
32	Oil pressure switch
33	Alternator
34	Front indicators
35	Left-hand front indicator
36	Right-hand front indicator
37	Headlamp
38	Parking lamp
39	Sensor
40	Speedometer lamp
41	Speedometer lamp
42	Speedometer
43	Oil pressure warning lamp
44	Headlamp failure indicator lamp
45	Left-hand indicator warning lamp
46	Right-hand indicator warning lamp
47	Neutral indicator lamp
48	Pilot lamp box
49	High beam indicator lamp
50	Tachometer lamp
51	Tachometer lamp
52	Tachometer
53	Front brake stop lamp switch

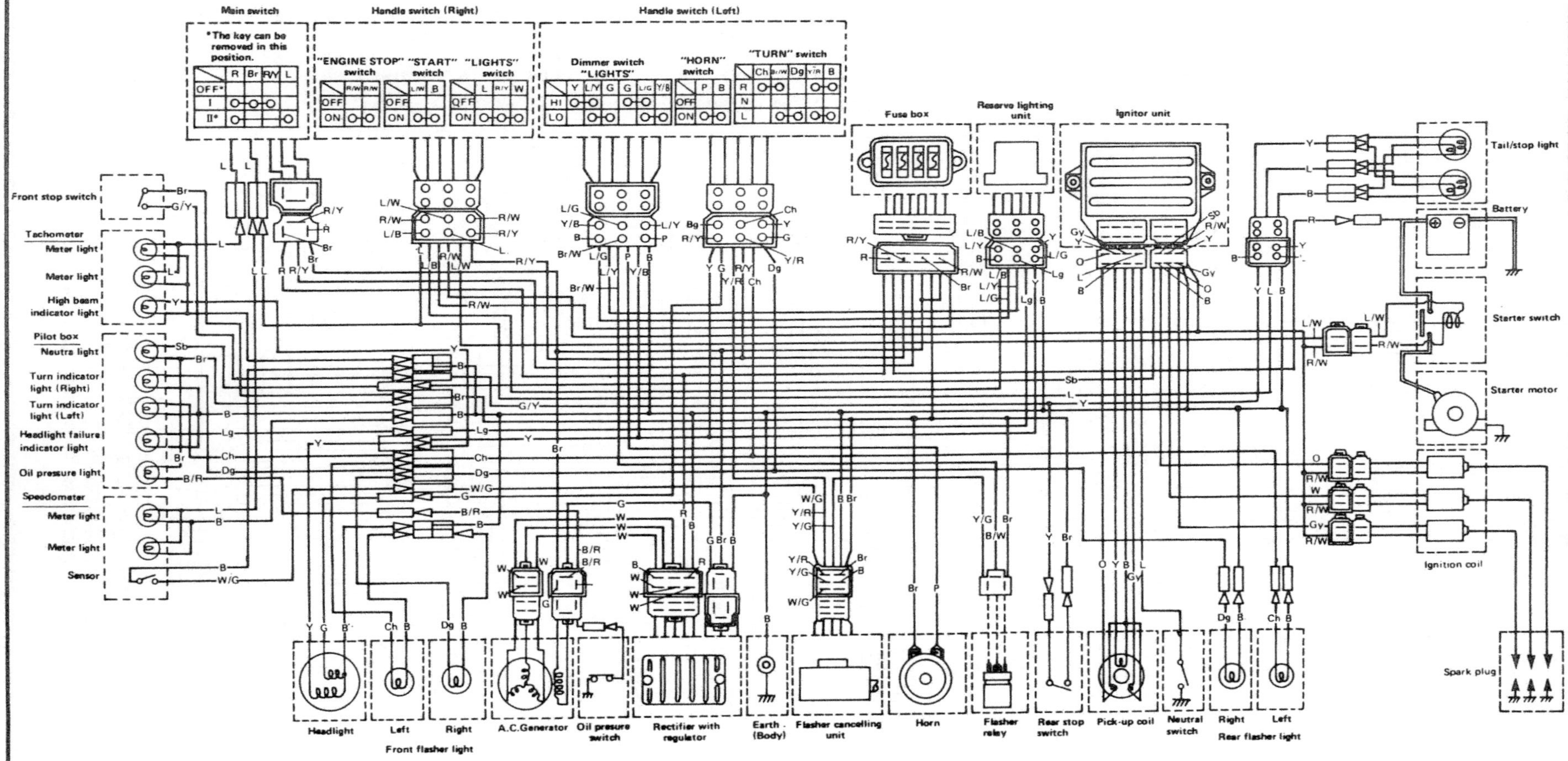

Wiring diagram for Yamaha XS750 E – US model

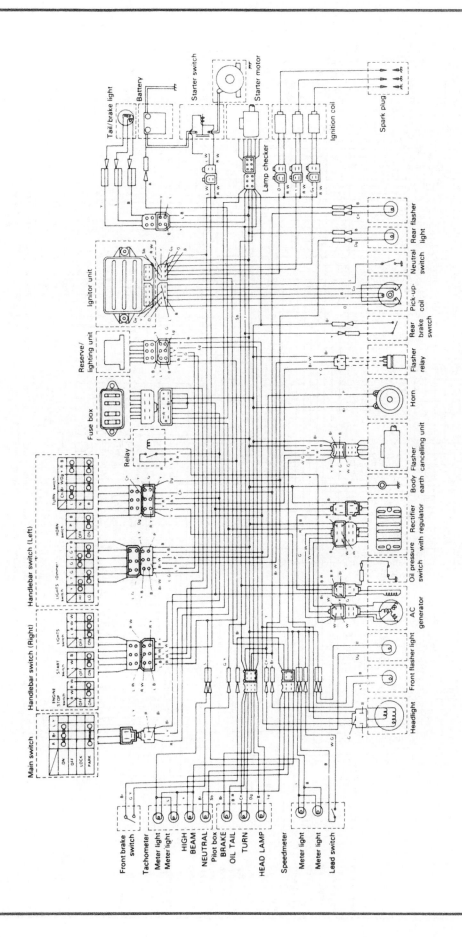

Wiring diagram for Yamaha XS750 SE and SF models

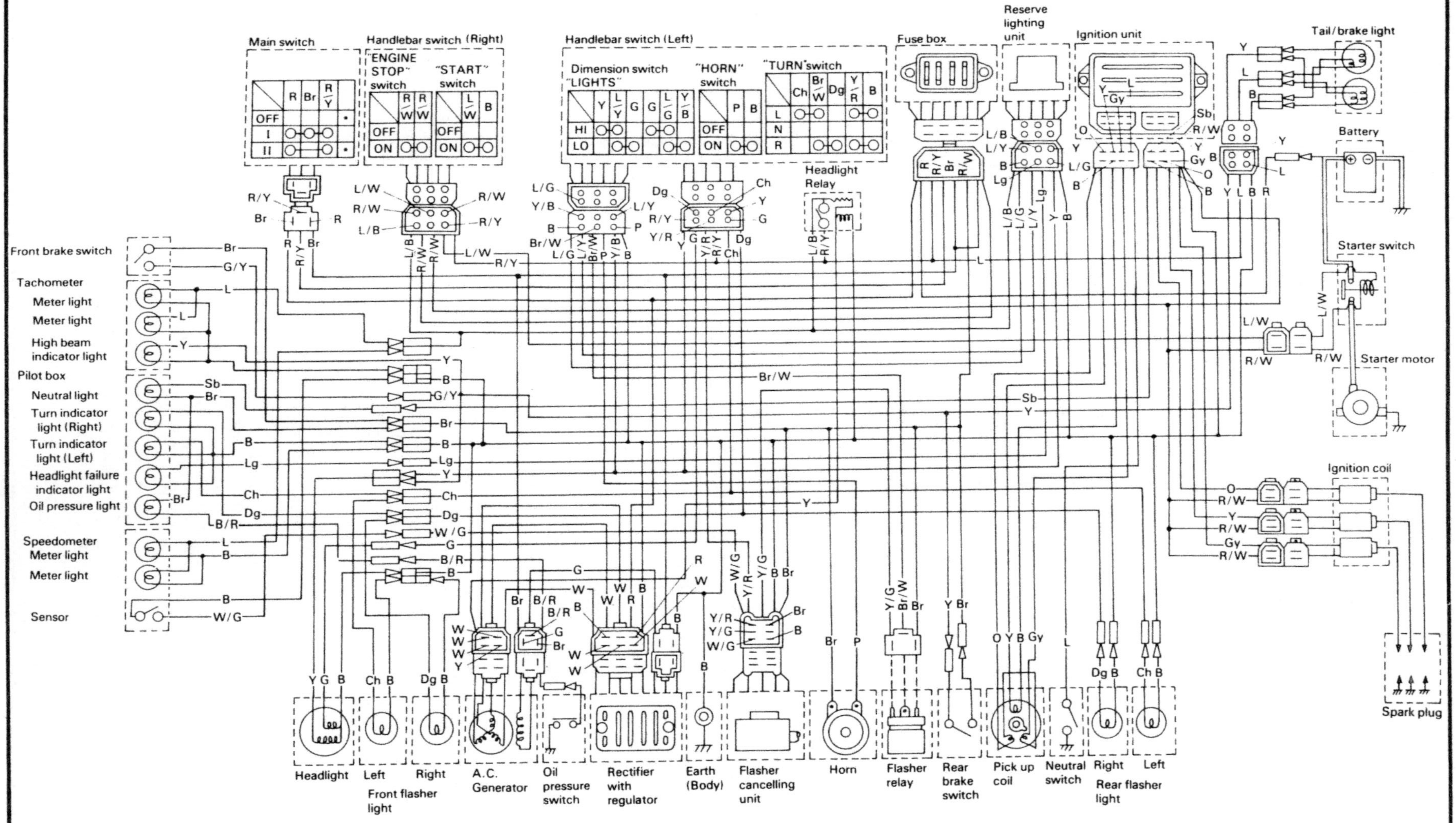

Tail/brake light

Battery

Starter switch

Starter motor

Ignition coil

Spark plug

Ignition unit

Reserve lighting unit

Right Left Rear flasher light

Neutral switch

Pick up coil

Rear brake switch

Fuse box

Flasher relay

"TURN" switch

"HORN" switch

Horn

Headlight Relay

Flasher cancelling unit

Handlebar switch (Left)

Dimension switch "LIGHTS"

Earth (Body)

Rectifier with regulator

Oil pressure switch

Handlebar switch (Right)

"START" switch

A.C. Generator

"ENGINE STOP" switch

Right Left Front flasher light

Main switch

Headlight

Front brake switch
Tachometer
Meter light
Meter light
High beam indicator light
Pilot box
Neutral light
Turn indicator light (Right)
Turn indicator light (Left)
Headlight failure indicator light
Oil pressure light
Speedometer
Meter light
Meter light
Sensor

Wiring diagram for Yamaha XS750 F model

The 1981 Yamaha XS850 G (UK) model

Engine unit of the XS850 G (UK)

Chapter 8 The XS850 G, SG, H, SH and LH models

Contents

Specifications

Specifications for the XS850 G, XS850 SG, XS850 H, XS850 LH and XS850 SH models are as stated in the main text of this manual except where listed below. Where no specific figures are given for the LH model refer to those for the SH model.

Specifications relating to Chapter 1

Engine
Bore	71.50 mm (2.815 in)
Capacity	826 cc (50.4 cu in)
Compression ratio	9.2 : 1

Cylinder barrel
Standard bore	71.50 mm (2.8150 in)
Service limit	71.60 mm (2.8189 in)

Piston and rings
Piston diameter:	
Standard	71.45 mm (2.8130 in)
1st oversize	71.75 mm (2.8248 in)
2nd oversize	72.00 mm (2.8346 in)
3rd oversize	72.25 mm (2.8445 in)
4th oversize	72.50 mm (2.8543 in)
Ring oversize (top and 2nd):	
1st oversize	0.25 mm (0.0098 in)
2nd oversize	0.50 mm (0.0197 in)
3rd oversize	0.75 mm (0.0295 in)
4th oversize	1.00 mm (0.0394 in)
Ring oversize (oil control):	
1st oversize	Brown
2nd oversize	Blue
3rd oversize	Black
4th oversize	Yellow
Ring end gap:	
Top and 2nd service limit	1.0 mm (0.0394 in)
Oil control ring	0.3 — 0.9 mm (0.0118 — 0.0035 in)
Service limit	1.5 mm (0.0591 in)
Ring side clearance:	
Top and 2nd service limit	0.15 mm (0.0059 in)

Valves and valve springs

Valve stem diameter:
  Inlet ... ...   6.975 — 6.990 mm (0.2746 — 0.2752 in)
  Exhaust ... ...   6.960 — 6.975 mm (0.2740 — 0.2746 in)
Valve guide/stem clearance:
  Inlet ... ...   0.010 — 0.040 mm (0.0004 — 0.0016 in)
    Service limit ... ...   0.10 mm (0.004 in)
  Exhaust ... ...   0.025 — 0.055 mm (0.0010 — 0.0022 in)
    Service limit ... ...   0.12 mm (0.005 in)
Valve clearance (engine cold):
  Inlet – UK models ... ...   0.16-0.20 mm (0.006-0.008 in)
  Inlet – US models ... ...   0.11-0.15 mm (0.004-0.006 in)
  Exhaust ... ...   0.21-0.25 mm (0.008-0.010 in)

Camshaft

Inlet:
  Standard lift ... ...   8.805 mm (0.347 in)
  Overall lobe height ... ...   36.805 ± 0.05 mm (1.449 ± 0.002 in)
  Service limit ... ...   36.65 mm (1.443 in)
  Journal diameter ... ...   28.341 ± 0.05 mm (1.116 ± 0.002 in)
  Service limit ... ...   28.19 mm (1.110 in)
Exhaust:
  Standard lift ... ...   8.305 mm (0.327 in)
  Overall lobe height ... ...   36.305 ± 0.05 mm (1.429 ± 0.002 in)
  Service limit ... ...   36.15 mm (1.423 in)
  Journal diameter ... ...   28.341 ± 0.05 mm (1.116 ± 0.002 in)
  Service limit ... ...   28.19 mm (1.110 in)
Runout service limit ... ...   0.1 mm (0.004 in)

Crankshaft

Main bearing clearance ... ...   0.060 — 0.082 mm (0.0024 — 0.0032 in)
Big-end bearing clearance ... ...   0.042 — 0.064 mm (0.0017 — 0.0025 in)

Clutch

Friction plate thickness ... ...   2.8 mm (0.11 in) — 7 off
  Service limit ... ...   2.6 mm (0.10 in)
Plain plate thickness ... ...   2.0 mm (0.079 in) — 4 off

Maximum warp limit ... ...   0.05 mm (0.0020 in)
Spring free length ... ...   42.8 mm (1.685 in)
  Service limit ... ...   41.8 mm (1.646 in)

Gearbox

Ratios (no of teeth):
  1st ... ...   2.285 : 1 (32/14)
  2nd ... ...   1.588 : 1 (27/17)
  3rd ... ...   1.300 : 1 (26/20)
  4th ... ...   1.095 : 1 (23/21)
  5th ... ...   0.956 : 1 (22/23)
Secondary drive spur gear ratio:
  XS850 G (UK) ... ...   1.062 : 1 (34/32)
  XS850 G (US) and XS850 H (US) ... ...   1.129 : 1 (35/31)
  XS850 SG (US), XS850 SH (US) & XS850 LH (US) ..   1.093 : 1 (35/32)
Middle bevel gear ratio ... ...   1.055 : 1 (19/18)
Final bevel gear ratio ... ...   2.909 : 1 (32/11)

Specifications relating to Chapter 2

Fuel tank capacity

XS850 G (UK) ... ...   24 litres (5.28/6.34 Imp/US galls)
All other models ... ...   17 litres (3.8/4.5 Imp/US galls)
Reserve ... ...   3 litres (0.66/0.8 Imp/US galls)

Carburettors

Make ... ...   Hitachi
Type:
  XS850 G (UK) HSC34 4E200
  XS850 G (US) and XS850 H (US) ... ...   3J3 — 00
  XS850 SG (US) and XS850 SH (US) ... ...   3J2 — 00
  XS850 LH (US) ... ...   HSC34 — 11
Main jet:
  XS850 G (UK)   148
  All other models   142

Jet needle:
    XS850 G (UK) ... ...   Y — 02
    All other models ... ...   Y — 01
Needle jet   ... ...   2.6 mm
Starter jet   40
Pilot jet   40
Float height ... ... 12.5 ± 0.5 mm (0.492 ± 0.020 in)
Fuel level:
    XS850 H (US) and XS850 SH (US) ... ... 1 ± 1mm (0.04 ± 0.04 in)
Pilot screw   ... ...   Preset
Main air jet   50
Pilot air jet   155
Throttle valve   ... ...   34 mm
Float valve seat   ... ...   2.0 mm
Engine idle speed   ... ...   1100 rpm

Oil capacity
Dry   ... ...   3.7 litres (6.5/7.8 Imp/US pints)
Oil change   ... ...   2.8 litres (4.9/5.9 Imp/US pints)

Specifications relating to Chapter 3

Ignition timing
Retarded   ... ... 10° BTDC at 1100 rpm
Advance starts   ... ...   2000 rpm
Full advance ...   . 34° BTDC at 4000 rpm

Pick-up coil resistance
XS850 H and XS850 SH   ... ...   650 ohms $\pm 20\%$ at 20°C (68°F)
All other models   ... ...   700 ohms $\pm 20\%$ at 20°C (68°F)

Ignition coil
Make ...   .   Hitachi
Type:
    XS850 H and XS850 SH   ... ...   CM11 — 55
    All other models   ... ...   CM11 — 54
Spark gap   ... ...   6 mm (0.24 in) at 500 rpm
Primary resistance   ... ...   2.75 ohms $\pm 10\%$ at 20°C (68°F)
Secondary resistance   ... ...   7.9 k ohms $\pm 20\%$ at 20°C (68°F)

Spark Plug
Make ...   .   NGK or Champion
Type ...   .   BP7ES or N — 7Y
Suppressor cap resistance:
    XS850 G (UK)   ... ...   5.5 k ohms
    All other models   ... ...   5.0 k ohms

Specifications relating to Chapter 4

Front forks
Oil capacity per leg:
    XS850 G (UK) ... ...   200 cc (7.04/6.76 Imp/US fl oz)
    XS850 G (US) and XS850 H (US)   ... ...   195 cc (6.86/6.59 Imp/US fl oz)
    XS850 SG (US) and XS850 SH (US) ...   .   275 cc (9.68/9.30 Imp/US fl oz)
Air pressure:
    Standard:
        XS850 H (US)   ... ... $14 - 21$ psi ($1.0 - 1.5$ kg/cm^2)
        XS850 G (US), XS850 SG (US) and XS850 SH (US)   ...   5.7 psi (0.4 kg/cm^2)
        Maximum...   ... ...   36 psi (2.5 kg/cm^2)
Spring free length:
    XS850 G (UK)   ... ...   502 mm (19.76 in)
    XS850 G (US) and XS850 H (US)   ... ...   508.5 (20.02 in)
    XS850 SG (US) and XS850 SH (US) ...   .   613 mm (24.13 in)

Rear suspension units
Spring free length:
    XS850 G (UK)   ... ...   237 mm (9.33 in)
    All other models   ... ...   243 mm (9.57 in)

Specifications relating to Chapter 5

Wheels
Type:
  Front ... ... MT 1.85 × 19 cast aluminium
  Rear:
    XS850 SG (US) and XS850 SH (US) ... ... MT 3.00 × 16 cast aluminium
    All other models ... ... MT 2.50 × 17 cast aluminium
Rim run-out:
  Horizontal ... ... 2.0 mm (0.079 in)
  Vertical ... ... 2.0 mm (0.079 in)

Brakes
Pad thickness service limit:
  Front:
    XS850 SG (US) and XS850 SH (US) ... ... 6.5 mm (0.26 in)
    All other models ... ... 6.0 mm (0.24 in)
  Rear ... ... 6.0 mm (0.24 in)
Disc thickness service limit:
  Front and rear ... ... 6.5 mm (0.26 in)
Disc maximum warpage:
  Front and rear ... ... 0.15 mm (0.006 in)

Tyres
Type:
  Front:
    XS850 G (UK) ... ... 3.25 H19 — 4PR tubed
    All other models ... ... 3.25 H19 — 4PR tubeless
  Rear:
    XS850 G (UK) ... ... 4.50 H17 — 4PR tubed
    XS850 G (US) and XS850 H (US) ... ... 4.50 H17 — 4PR tubeless
    XS850 SG (US), XS850 SH (US) and XS850 LH (US) ... 130/90 — 16 67H tubeless

Pressures (cold):

	Front	Rear
Up to 198 lb (90 kg) load:		
XS850 H (US)	40 psi (2.8 kg/cm^2)	40 psi (2.8 kg/cm^2)
All other models	26 psi (1.8 kg/cm^2)	28 psi (2.0 kg/cm^2)
198 lb (90 kg) to 406 lb (184 kg) load:		
XS850 H (US)	40 psi (2.8 kg/cm^2)	40 psi (2.8 kg/cm^2)
All other models	28 psi (2.0 kg/cm^2)	32 psi (2.3 kg/cm^2)
High speed riding:		
XS850 H (US) and XS850 G (UK)	40 psi (2.8 kg/cm^2)	40 psi (2.8 kg/cm^2)
All other models	28 psi (2.0 kg/cm^2)	32 psi (2.3 kg/cm^2)
198 lb (90 kg) to 410 lb (186 kg) load and high speed riding:		
XS850 SG (US) and XS850 SH (US)	28 psi (2.0 kg/cm^2)	32 psi (2.3 kg/cm^2)

Advised minimum tread depth ... ... 0.8 mm (0.03 in)

Specifications relating to Chapter 6

Alternator
Make ... ... Hitachi
Type ... ... LD 120
Output ... ... 14V/20A (280 W) at 5000 rpm
Field coil resistance ... ... 3.5 ohms ± 10% at 20°C (68°F)
Stator coil resistance ... ... 0.4 ohm ± 10% at 20°C (68°F)

Starter motor
Field coil resistance ... ... 0.01 ohm at 20°C (68°F)
Armature coil resistance ... ... 0.007 ohm at 20°C (68°F)
Starter relay switch:
  Type ... ... Hitachi A104 — 70
  Winding resistance ... ... 3.5 ohms at 20°C (68°F)
  Cut-in voltage ... ... 6.5 volts
Voltage control regulator:
  Type ... ... National RD 1143 or Shindengen SH 233
  Regulated voltage ... ... 14.5 ± 0.3 volts
  Amperage allowed ... ... 4 amps

Bulbs
Headlamp ... ... 60/55 W
Tail/stop lamp:
  XS850 G (UK) ... ... 5/21 W
Number plate lamp:
  XS850 SG (US) and XS850 SH (US) ... ... 3.8 W

Model dimensions

	XS850 G (UK)	XS850 G and H (US)	XS850 H (US) Touring	XS850 SG SH and LH
Height	1120 mm (44.1 in)	1190 mm (46.9 in)	1570 mm (61.8 in)	1265 mm (49.8 in)
Width	740 mm (29.1 in)	900 mm (35.4 in)	935 mm (36.8 in)	925 mm (36.4 in)
Length	2155 mm (84.8 in)	2175 mm (85.6 in)	2270 mm (89.4 in)	2200 mm (86.6 in)
Wheel base	1465 mm (57.7 in)	1450 mm (57.1 in)	1450 mm (57.1 in)	1490 mm (58.7 in)
Seat height	815 mm (32.1 in)	815 mm (32.1 in)	815 mm (32.1 in)	800 mm (31.5 in)
Ground clearance (minimum)	130 mm (5.1 in)	140 mm (5.5 in)	140 mm (5.5 in)	150 mm (5.9 in)
Weight	236 kg (520 lb)	241 kg (531 lb)	274 kg (604 lb)	237 kg (522 lb)

Torque wrench settings
In addition to those given in the main text.

	lbf ft	kgf m
Cylinder head bolt (8 mm)	14.5	2.0
Camshaft sprocket	36.2	5.0
Camshaft chain tensioner nut (8 mm)	6.5	0.9
Camshaft chain tensioner bolt (6 mm)	7.2	1.0
Starter clutch bolt (8 mm)	21.7	3.0
Alternator rotor bolt	36.2	5.0
Engine oil drain plug	31.0	4.3
Middle gear oil drain plug	31.0	4.3
Oil filter bolt	23.1	3.2
Crankcase/cylinder head oil delivery pipe	14.5	2.0
Oil pressure switch	13.0	1.8
Oil pump gear retaining nut	5.8	0.8
Oil pump assembly retaining bolts	7.2	1.0
Oil strainer cover retaining bolts	7.2	1.0
Strainer cover baffle plate retaining screws	5.8	0.8
Crankcase breather cover retaining bolts	7.2	1.0
Clutch retaining nut	79.6	11.0
Clutch adjusting screw locknut	14.5	2.0
Clutch spring screw	6.5	0.9
Gearchange pedal bolt	7.2	1.0
Neutral indicator switch	14.5	2.0
Exhaust pipe/cylinder head nuts (8 mm)	14.5	2.0
Exhaust pipe/silencer clamp bolts (6 mm)	7.2	1.0
Exhaust pipe balancer pipe clamp bolt (8 mm)	14.5	2.0
Front lower engine mounting nut (10 mm)	39.8	5.5
Rear lower engine mounting nut (12 mm):		
XS850 (UK)	68.7	9.5
All other models	18.1	2.5
Front lower engine mounting nut (8 mm)	14.5	2.0
Front fork upper yoke pinch bolt nut (8 mm)	10.8	1.5
Upper yoke/steering stem bolt (14 mm)	39.1	5.4
Upper yoke/fork leg pinch bolt nuts (8 mm)	10.8	1.5
Handlebar retainer/upper yoke nut (8 mm)	13.0	1.8
Front fork lower yoke/fork leg pinch bolts (8 mm)	14.5	2.0
Rear suspension unit/swinging arm bolt	28.2	3.9
Rear suspension unit/final drive casing nut	28.2	3.9
Rear suspension units/frame nuts	21.7	3.0
Swinging arm pivot shafts	3.95	0.55
Pivot shaft locknuts	72.3	10.0
Swinging arm/final drive casing nuts	30.4	4.2
Final drive casing oil drain bolt	31.1	4.3
Front wheel spindle retaining nut	77.4	10.7
Spindle holder retaining nuts:		
XS850 G and XS850 H	14.5	2.0
Front fork/wheel spindle pinch bolt:		
XS850 SG and XS850 SH	14.5	2.0

Rear wheel spindle retaining nut	...	...	...	...	108.5	15.0
Swinging arm/wheel spindle pinch bolt	...	...	...	...	4.5	0.6
Brake hose connections ...	...	...	...	...	18.8	2.6
Front brake disc/wheel hub retaining bolts	...	...	...	...	14.5	2.0
Brake caliper/front fork retaining bolts	...	...	...	...	18.8	2.6
Rear brake disc/wheel hub retaining bolts	...	...	...	...	14.5	2.0
Rear brake caliper/support bracket retaining nut	...	...	...	13.0	1.8
Rear brake master cylinder/frame retaining bolt	...	...	...	16.6	2.3
Brake caliper bleed screws	...	...	...	...	4.3	0.6

1 Introduction

The information contained in the following Sections covers all XS850 models introduced in the UK and US for the years 1980 (model suffix G) and 1981 (model suffix H). All of these models are similar in many aspects to the XS750 models covered in Chapter 1 to 7 of this manual, many of the improvements made on the later XS750 models, and noted in Chapter 7, being carried over to the XS850 models. It is therefore most important that the contents of both this Chapter and of Chapter 7 are fully consulted before reverting to the main text for further service information. Full comparison should be made between the item fitted to the particular model type being serviced and the figures included in each Chapter to ensure that the item identifies correctly with that covered in the text.

Apart from the obvious change in engine capacity and the subsequent changes in engine specifications, the power unit and frame design of the XS850 models remain similar to those used for the XS750 models. All the design changes included in this Chapter have been made with a view to improving engine efficiency, both in terms of reliability and ease of maintenance, and to giving both improved handling and roadholding qualities to the machine.

Examples of these design changes include the fitting of an oil cooler system, a simplified electronic ignition system (H suffix models only) and a redesigned exhaust system. Hitachi carburettors have been introduced for the first time by Yamaha on the XS850 models and form a break with their traditional use of Mikunis. The positioning and layout of these carburettors remains similar to that used on the XS750 models, the same design of air cleaner assembly as described in Chapter 7 being retained. Improvements to the suspension from that used on later XS750 models include the introduction of air assisted front forks and of rear suspension units which are adjustable both in the damping and spring pre-load settings. It should be noted, however, that the XS850 G model for the UK market still retains the type of suspension used on the later XS750 models. At the time of inclusion of this updated information comprehensive specifications for the XS850 LH Midnight Special model were not available. For all practical purposes, however, the LH is similar to the SH model and the SH information should be referred to as a general guide.

2 Gearbox components: removal, examination, renovation and refitting

1 When carrying out the above listed servicing procedures on the gearbox components fitted to the XS850 models, refer to the appropriate Section in Chapter 1 whilst noting the detail differences between the earlier and later gearbox assemblies shown on the figure accompanying this text. Refer also to Section 4 of Chapter 7 when refitting the middle gear bearing.

3 Checking the primary driven gear thrust play

1 On completion of fitting the primary driven gear, the primary drive chain and the clutch damper assembly, retain the driven gear in position with the circlip and measure the amount of thrust play. If the amount of play exceeds 0.3 mm (0.012 in), an adjusting shim of 0.3 mm or 0.5 mm should be inserted between the clutch housing and primary driven gear surfaces as shown in the figure accompanying this text.

4 Clutch: checking the clutch housing thrust play

1 On completion of fitting the clutch housing and retaining it in position with the plate washer and circlip, check that the amount of thrust play is no greater than 0.2 mm (0.008 in). If the amount of play is found to be excessive, it should be decreased by the addition of one or two shims between the plate washer and clutch housing as shown in the figure accompanying this text.

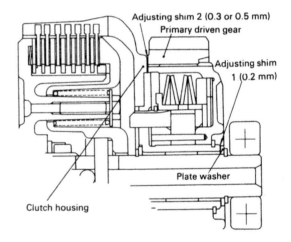

Adjusting shim 2 (0.3 or 0.5 mm)
Primary driven gear
Adjusting shim 1 (0.2 mm)
Plate washer
Clutch housing

Fig. 8.1. Adjusting the clutch housing and primary driven gear thrust play

5 Carburettors: removal and refitting

1 Following the procedures given in the main text, remove the seat and fuel tank from their frame attachment points. Unclip the two side panels and place them, together with the seat and tank, in a safe storage space.

2 Disconnect the air cleaner casing from its frame attachment points by removing the two bolts passing through each of the frame downtube brackets located either side of the casing. Loosen the three carburettor to air cleaner casing retaining clips and pull the casing rearwards to clear the carburettor inlets. Loosen the three carburettor to cylinder head stub retaining clips and pull the complete carburettor assembly rearwards to clear the stub ends. Some difficulty may be experienced in separating the carburettor assembly from both the air cleaner casing and the cylinder head stubs. In practice it was found that the aid of an assistant working from the opposite side of the machine was invaluable, if not vital, in steadying the carburettor assembly whilst it was freed.

3 Check that all pipes have been freed from their attachment clips on the carburettor assembly and pull the breather pipe off the air cleaner casing stub before attempting to move the carburettor assembly sideways (to the left) to expose the throttle cable linkage. Again some difficulty was experienced in easing the carburettor mouths past their attachment points, but with some care and patience the task was eventually accomplished. Once fully exposed, the throttle cable can be disconnected from its attachment point on the carburettor linkage and the complete carburettor assembly then lifted clear of the machine.

4 Refitting the carburettor assembly is essentially a reversal of the removal procedure, noting the following points. Lightly smear the mouths of the air cleaner casing and cylinder head stubs with a small amount of soapy liquid; this will greatly reduce the amount of effort needed to push the carburettor mouths into their respective locations. Again it was found in practice that the aid of an assistant was desirable when holding back the air cleaner casing and retaining the clips on both the casing and cylinder head stubs in order to allow the carburettor assembly to be located in position. The carburettor assembly will also need to be held steady whilst the air cleaner casing is pushed forward onto the carburettor mouths. Finally, with the fuel tank refitted and all the fuel pipes and breather pipes correctly routed and clipped in position, carry out a final check for fuel leaks around all disturbed connections.

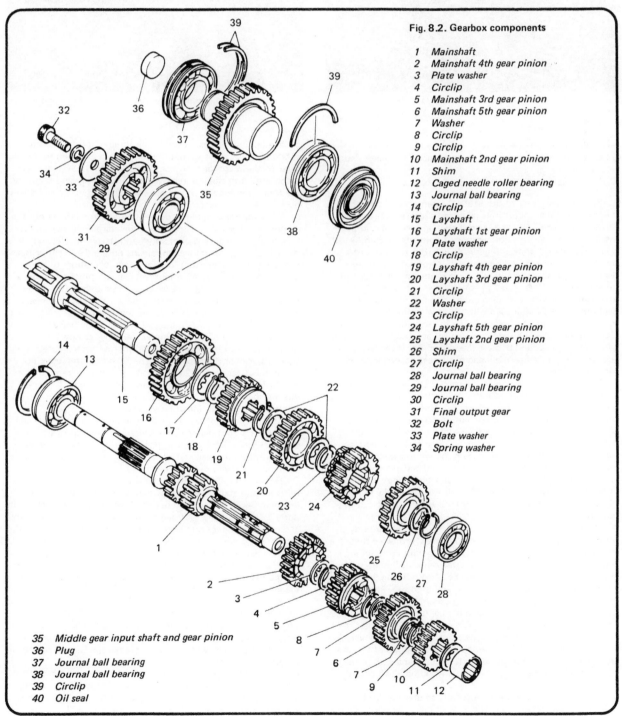

Fig. 8.2. Gearbox components

1   Mainshaft
2   Mainshaft 4th gear pinion
3   Plate washer
4   Circlip
5   Mainshaft 3rd gear pinion
6   Mainshaft 5th gear pinion
7   Washer
8   Circlip
9   Circlip
10 Mainshaft 2nd gear pinion
11 Shim
12 Caged needle roller bearing
13 Journal ball bearing
14 Circlip
15 Layshaft
16 Layshaft 1st gear pinion
17 Plate washer
18 Circlip
19 Layshaft 4th gear pinion
20 Layshaft 3rd gear pinion
21 Circlip
22 Washer
23 Circlip
24 Layshaft 5th gear pinion
25 Layshaft 2nd gear pinion
26 Shim
27 Circlip
28 Journal ball bearing
29 Journal ball bearing
30 Circlip
31 Final output gear
32 Bolt
33 Plate washer
34 Spring washer

35 Middle gear input shaft and gear pinion
36 Plug
37 Journal ball bearing
38 Journal ball bearing
39 Circlip
40 Oil seal

5.3 Move the carburettor assembly sideways to expose the throttle cable linkage

5.4 Ensure that all carburettor connections are fully tightened

6 Carburettors: dismantling, examination, cleaning and reassembly

1   Before attempting to dismantle the carburettor assembly, it should be noted that it is inadvisable to separate the individual carburettors from the two bracing bars that hold the complete assembly together. It is not necessary to disturb the carburettor body to bracing bar securing screws in order to carry out a normal servicing procedure on each carburettor and any disturbance of these screws will result in the assembly becoming misaligned, resulting in the assembly having to be realigned in accordance with the instructions given in the following Section of this Chapter. Note also that when removing any screw connections, it is essential that the screwdriver used is as close fitting as possible. This will prevent any damage occurring to the screw heads and because the screwdriver is less likely to slip, greatly reduce the likelihood of damage occurring to the carburettor body.

2   Commence by positioning the carburettor assembly, float chambers uppermost, on a clean work surface. Proceed to service separately each carburettor as follows. Unscrew and remove the four float chamber body securing screws and lift the chamber body to expose the float and jet assemblies located beneath it. Remove the jet retaining plate and free the float assembly from the carburettor body by withdrawing its pivot pin. Note that this pin is splined at one end and can only be removed by tapping it out from the plain end using a small drift. Support the column retaining the splined end of the pin to prevent it being bent or broken should the pin be seized in position.

3   Remove the needle valve seat from the carburettor body by unscrewing its retaining plate securing screw, removing the plate and pulling the valve seat from its location. Pull the main jet, pilot jet and starter jet out of the three central columns, noting the position of each one for reference when refitting. Invert the carburettor and catch the needle jet, emulsion tube and starter air jet as they fall from their housings, again noting the position of each item for reference when refitting. It may be necessary to tap sharply the carburettor into a cupped hand in order to free the jets if they have become stuck within their housings over a long period of time. An alternative method of freeing the needle jet is to push down on its exposed end within the carburettor bore, using a piece of soft wood such as a clothes peg.

4   Reposition the carburettor assembly so that the diaphragm covers are uppermost and unclip each of the three starter plunger

bush retaining E-rings. Remove the three nylon bushes and rotate the lever shaft so that the three levers attached to it move clear of the diaphragm covers. The diaphragm cover of each carburettor can now be removed by unscrewing and removing the four retaining screws. Remove the diaphragm spring followed by the diaphragm assembly.

5   The only remaining removable components left within the carburettor body at this point are the starter plunger and its spring; the main air jet, pilot air jet and starter nozzle; and the air vent orifices. Of these components only the air vent orifices can be removed without resorting to separation of the carburettor body halves. Should the starter plunger assembly be suspected of being unserviceable, the carburettor body halves will need to be separated, which in turn will mean separating the carburettor from the two bracing bars. Note that two of the screws securing the carburettor body halves together are located within the diaphragm housing; removal of the diaphragm from its housing is therefore necessary before the body halves can be separated. The manufacturer does not recommend separation of the body halves and thus, unless absolutely necessary, do not disturb the casings. If the air jets are merely thought to be blocked cleaning can be carried out without dismantling using a low pressure air supply.

6   Before commencing the cleaning and examination procedure detailed in the following paragraphs, cover an area of the work surface with a piece of clean paper or rag. This will prevent any cleaned components placed upon it from becoming contaminated with dirt or moisture.

7   Remove the O-rings from the needle valve seat, main jet, pilot jet and starter jet and fit new items after the seat and jets have been cleaned. Remove the float chamber to carburettor body sealing gasket and fit a new item during reassembly of the carburettor. Inspect the diaphragm assembly for damage to the rubber and piston. If the rubber is torn or the piston scratched, then the jet needle assembly must be removed from the piston and the diaphragm assembly replaced with a new item. The jet needle must also be removed for renewal if inspection with a magnifying glass shows it to be worn. To remove the needle from the piston, grip the piston firmly in one hand and unscrew the needle retaining cap from within the centre of the piston. Withdraw the cap, spring, plain washer and the needle together with its nylon bush and E-ring. Note that in no circumstances should the jet needle set screw, located at the base of the piston, be disturbed from its setting.

8   Inspect the needle valve and needle valve seat for signs of wear and damage on the seating faces. The needle and seat should be renewed as a pair otherwise imperfect seating between the two components will persist.

9 Inspect all removed springs for signs of damage or corrosion and renew as necessary. Check that the twin floats are in good order and are not punctured. It is not possible to effect a permanent repair. In consequence, a new replacement should always be fitted if damage is found. Finally, check that all mating surfaces on the carburettor body are flat by using a straight-edge laid across the mating surfaces.

10 Clean each carburettor component carefully in clean fuel, using a fine-bristled nylon brush (a toothbrush is ideal) to remove any stubborn traces of sediment. Shake any residue of fuel from the cleaned component and use a jet of compressed air to blow any remaining sediment or fuel from any internal passageways or jet orifices. Take care when using compressed air to ensure that the blast of air is directed away from one's person; it is advisable to wear eyeshields during this operation in order to prevent any fuel entering the eyes. If a compressed air supply is not available a blast of air from a tyre pump will usually suffice.

11 Never use a piece of wire or any pointed metal object to clear a blocked jet. It is only too easy to enlarge a jet in these circum-

stances and thus upset the carburettor setting. If a blast of compressed air fails to remove the blockage, a bristle removed from a fine nylon brush may be used but only as a last resort. Always avoid using a piece of rag to wipe components because there is always a risk of particles of lint becoming lodged in the jet orifices.

12 Reassembly of the carburettors is essentially a reversal of the dismantling procedure, noting the following points. Never use excessive force when reassembling the component parts because it is easy to shear a jet or one of the smaller screws. Furthermore, the carburettor is cast in a zinc based alloy which itself does not have a high tensile strength. If any of the castings are damaged during reassembly, they will almost certainly have to be renewed.

13 When refitting the diaphragm assembly, ensure that the needle passes cleanly into its location within the needle jet by viewing it through the carburettor bore. Ensure also that the tab projecting from the periphery of the diaphragm is correctly located in the matching recess in the carburettor body before fitting the diaphragm cover.

6.2a Remove the float chamber body to expose the float and jet assemblies

6.2b Note the float pivot pin is splined at one end

6.3a Unscrew the needle valve seat retaining plate securing screw ...

6.3b ... to allow the valve seat to be removed

6.3c Pull the pilot jet from its retaining column ...

6.3d ... followed by the emulsion tube

6.4a Remove each of the three nylon bushes ...

6.4b ... and rotate the lever shaft clear of the diaphragm covers

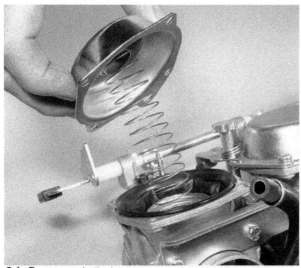

6.4c Remove each diaphragm cover and remove the spring ...

6.4d ... and diaphragm/piston assembly located beneath it

6.5 Note that two of the carburettor body securing screws are located within the piston diaphragm housing

6.7a Unscrew the needle retaining cap from the centre of the diaphragm piston ...

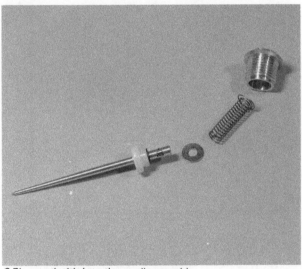

6.7b ... and withdraw the needle assembly

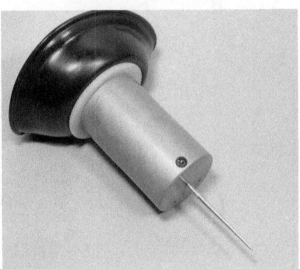

6.7c Do not disturb the jet needle set screw from its setting

6.8 Inspect the needle valve seating face for wear and damage

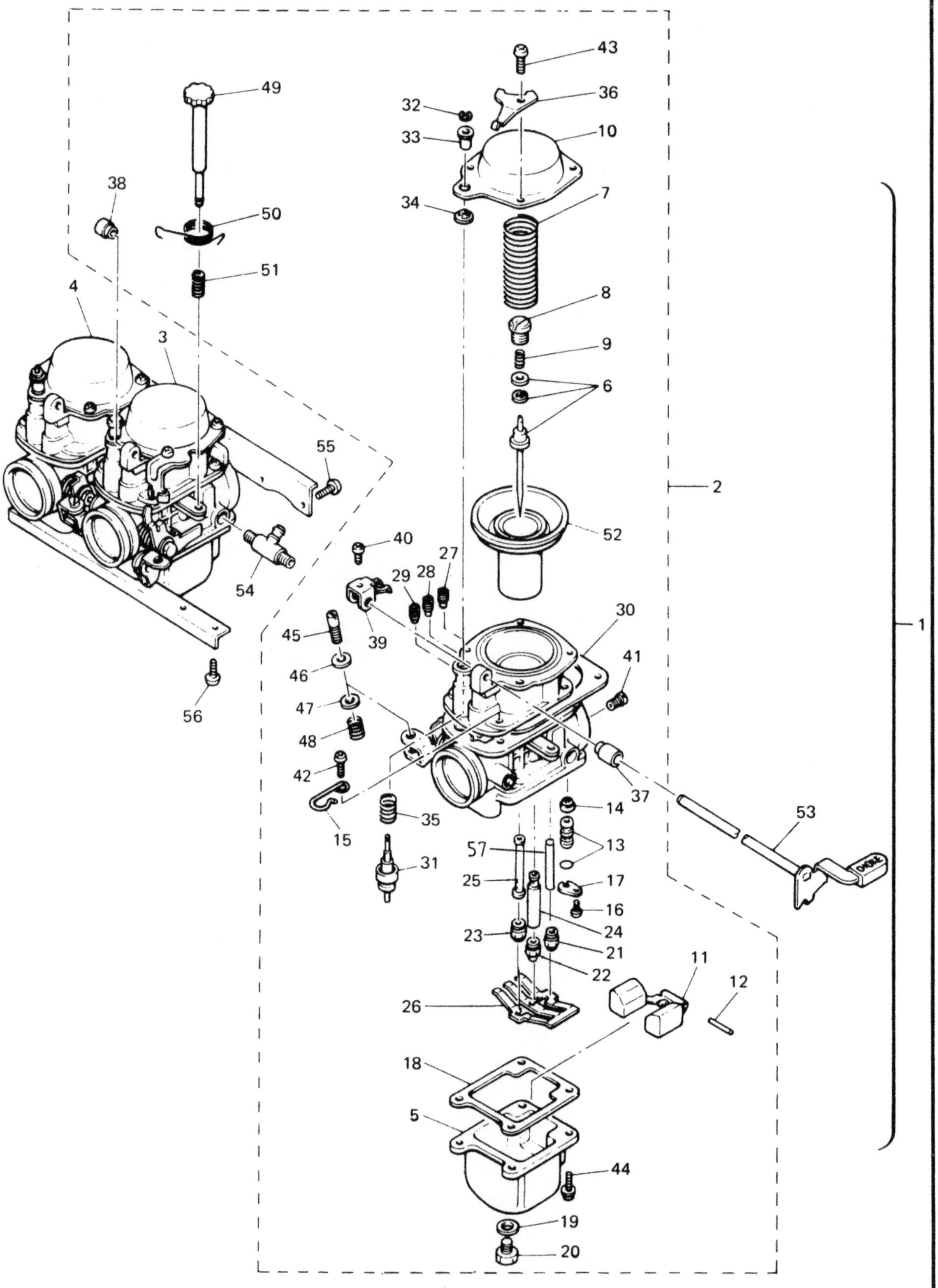
43
49
32
36
33
10
38
34
7
50
51
4
8
3
9
6
55
2
52
40
27
29
28
54
30
1
45
39
41
46
56
47
48
42
37
14
53
35
15
57
13
31
25
17
16
24
23
21
22
11
12
26
18
5
44
19
20

6.12 The carburettor assembly, showing the throttle and choke linkage assemblies

Fig. 8.4. Location of the two carburettor air orifices

Fig. 8.3. Carburettors — component parts

1	Carburettor assembly	28	Pilot air jet
2	No 1 carburettor assembly	29	Starter nozzle
3	No 2 carburettor assembly	30	Gasket
		31	Starter plunger
4	No 3 carburettor assembly	32	E-ring
		33	Bush
5	Float chamber body	34	Plunger cap cover
6	Jet needle assembly	35	Plunger spring
7	Diaphragm spring	36	Starter lever spring
8	Cap	37	Bush
9	Spring	38	Bush
10	Diaphragm cover	39	Starter-lever
11	Float assembly	40	Screw
12	Float pin	41	Air vent orifice
13	Needle valve assembly	42	Screw
14	Filter	43	Screw
15	Clip	44	Screw
16	Screw	45	Synchronizing screw
17	Plate	46	Washer
18	Gasket	47	Washer
19	Washer	48	Throttle stop spring
20	Screw	49	Throttle stop screw
21	Main jet	50	Spring
22	Pilot jet	51	Throttle stop spring
23	Starter jet	52	Diaphragm assembly
24	Emulsion tube	53	Starter lever shaft
25	Starter air jet	54	Union
26	Jet retaining plate	55	Screw
27	Main air jet	56	Screw
		57	Needle jet

7 Carburettor assembly: realignment

1   If, during the servicing procedures listed in the previous Sections, the individual carburettors have been detached from the two bracing bars, or even if the bar securing screws have been loosened in error, the carburettor assembly must be realigned by using the following procedure.

2   Place each carburettor, with the throttle linkage side uppermost, on a flat surface plate. A sheet of plate glass is the ideal substitute for a proper machined surface plate. Check that the plate surface beneath the carburettors is scrupulously clean and that the carburettor mouths are pressed firmly down on the surface. Place the first of the bracing bars in position across the top of the float chambers and secure it to the carburettor body attachment points with the six screws. Note that each screw should have a spring washer fitted beneath its head and a thread locking compound applied to its threaded portion before being fitted in position. Keep an even pressure on all three carburettors whilst tightening the screws to ensure correct alignment; a piece of planed wood placed across the upward facing carburettor mouths will aid in achieving this.

3 With the first bracing bar secured in position, the carburettor assembly can be inverted and the second bar fitted. Again, the securing screws should have spring washers fitted and a locking compound applied to their threads. There is no need to apply pressure to the assembly, as in the case of the first bar, when tightening these screws, but a final check should be carried out on completion to ensure that alignment has in no way been disturbed.

8 Carburettor settings

1 As stated in the main text, all jet sizes, etc are predetermined by the manufacturer and should not require changing during normal running of the machine. Raising or lowering of the jet needle can still be achieved by changing the position of the needle clip, although the design of the needle seat differs from that shown in the main text.

2 Measurement of the float height must be carried out with the carburettor held upside down. The tang between the two float arms should be just touching the float needle and the measurement taken from the top of the float to the float chamber gasket base surface (gasket removed). The distance measured should be 12.5 ± 0.5 mm (0.492 ± 0.020 in). If incorrect, the float height may be adjusted by bending the tang whilst ensuring that both float arms are kept parallel to each other.

3 The XS850 H and SH models have a nozzle cast in the base of the float chamber body to which a clear vinyl pipe can be attached to measure the fuel level within the float chamber. To ensure a correct fit over the float chamber nozzle, the inside diameter of the pipe should be as near to 6 mm (0.24 in) as possible. Measurement of the fuel level using the following procedure will obviate any need for the float chamber body to be removed in order to check the float height should problems of fuel flooding or starvation be encountered.

4 Before checking the fuel level, position the machine on its main stand on a level surface. Tilt the machine on the stand so the carburettors are positioned vertically and block it in position by placing a jack or wooden block beneath the engine sump. Connect the pipe to the nozzle of the left-hand carburettor, position it alongside the carburettor as shown in the accompanying figure and loosen the float chamber drain screw. Start the engine and allow it to idle for a few minutes in order to allow the fuel to find its correct level. Check that the fuel tap is in the ON or RES position before starting the engine.

5 Note the fuel level in the pipe in relation to a point on the carburettor body and leaving it attached to the nozzle of the left-hand carburettor, reposition the pipe alongside the right-hand carburettor. The fuel level in the pipe should align with a point on the body of the right-hand carburettor which corresponds with that point noted on the left-hand carburettor. If these two points correspond then the carburettor assembly is positioned in a horizontal plane along its length. This is essential to ensure the accurate checking of the fuel level. If the two points do not correspond, the position of the main stand must be altered by placing strips of wood beneath one of its feet until the carburettor assembly is set in the required horizontal plane.

6 With the carburettors correctly positioned, check the fuel level in each carburettor by repeating the procedure given in paragraph 4. The fuel level should settle in the pipe 1 ± 1 mm (0.04 ± 0.04 in) above the float chamber body to carburettor body mating surface. If the level is found to be incorrect, the carburettor assembly must be removed from the machine and the float assembly inspected for damage. Detach the needle valve from the float tang and inspect its seating face and the seating face of the needle valve seat in the carburettor body for damage or wear. Finally, check the float height as detailed in paragraph 2.

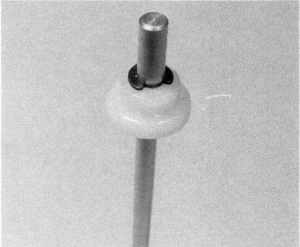

8.1 Change the clip position to raise or lower the jet needle

8.2a Measure the float height as shown ...

8.2b ... and carry out any necessary adjustment by bending the float tang (arrowed)

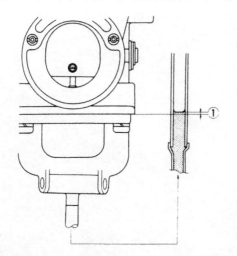

Fig. 8.5. Measuring the carburettor fuel level — XS850 H and SH

1   Fuel level

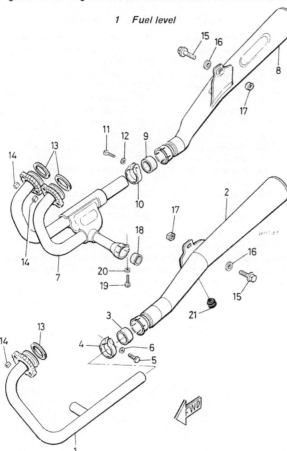

Fig. 8.6. Exhaust system assembly — XS850 G (UK)

1	Exhaust pipe	12	Plate washer
2	Silencer	13	Gasket
3	Gasket	14	Nut
4	Clamp	15	Flange bolt
5	Bolt	16	Plate washer
6	Plate washer	17	Nut
7	Exhaust pipe assembly	18	Gasket
8	Silencer	19	Bolt
9	Gasket	20	Plate washer
10	Clamp	21	Main stand stop rubber
11	Bolt		

9   Exhaust system: removal and refitting

1   Removal of the exhaust system is a simple and straight-forward procedure which can be simplified even further with the aid of an assistant. Commence removal by unscrewing and removing the exhaust pipe to cylinder head retaining screws. Loosen the clamp bolt at the balancer pipe connection located beneath the engine. With an assistant positioned one side of the machine and supporting the system so as to prevent it pivoting forward, remove the nut from each of the silencer to frame mounting attachment bolts. Support the system from both sides of the machine and withdraw the silencer attachment bolts. The complete system may now be lowered away from its mounting points and the system halves separated by pulling apart the balancer pipe connection. Take great care, when lowering the pipe to cylinder head connections past the oil cooler feed and return pipes, to avoid snagging the braided section of the pipe.
2   Fitting the system can be achieved by reversing the removal procedure. Always fit new gaskets to the pipe to cylinder head connections and tighten all disturbed connections to the figures given in the Specifications at the beginning of this Chapter. It should be noted that when removing or refitting the system, it should never be allowed to hang from the cylinder head connections because the weight of the system will place an unacceptable strain on them.
3   The XS850 G models have detachable silencers which may be separated from the exhaust pipe ends with the pipe assembly left in position on the machine. To remove each silencer, loosen the pipe to silencer clamp bolt, remove the silencer to frame attachment bolt and pull the silencer rearwards whilst twisting it slightly to free it from the pipe end.

10 Oil cooler assembly: general maintenance, removal and refitting

1   The oil cooler matrix on all XS850 models is mounted on the frame front downtubes to gain the best possible benefit from the cooling airflow. The feed and return pipes are interconnected at their lower ends with a distributor block interposed between the oil filter chamber and sump.
2   To maintain peak efficiency the matrix should be kept clear of any debris, preferably using an air jet directed from behind to blow out the air channels. Avoid using sharp instruments to dislodge any foreign matter; this may easily lead to damaged vanes. Should leakage of the matrix occur renewal of the complete component will probably be the only satisfactory solution. Repair is unlikely to be successful.
3   To remove an unserviceable matrix, separate the two pipes from their unions at the base of the unit by unscrewing the large retaining cap nuts. Note that the unions must be prevented from turning by placing an open-ended spanner across their flats. Support the matrix and remove the four forward facing bolts to release it from the frame mounting points. Fitting a serviceable matrix is a reversal of the removal procedure, taking great care not to overtighten the union cap nuts. Check that the union dowel pins are in good condition and that the O-rings are renewed and wiped with clean engine oil before fitting.
4   If the feed pipe or return pipe are found to be damaged or are seen to be leaking at their unions to the matrix or distributor block, they can be removed by first unscrewing the pipe to frame retaining clip bolt. Release the pipe from the distributor block by removing the two allen-headed bolts and release the pipe to matrix union as described above. Any leaks at the pipe unions are usually due to deteriorating O-rings; these should be renewed if at all suspect. Check the condition of the rubber protective tube between the hose and its frame retaining clips and renew it if it shows signs of wearing through. Any damage to the pipes must be thoroughly investigated because any leakage of oil could well lead to disastrous consequences.

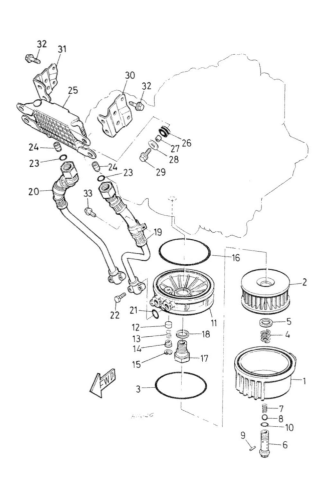

10.3 Release the oil cooler matrix from its two mounting points

Fig. 8.7. Oil cooler/cleaner assembly

1	Oil filter cover	19	Left-hand hose assembly
2	Filter element	20	Right-hand hose assembly
3	O-ring	21	O-ring
4	Spring	22	Bolt
5	Plate washer	23	O-ring
6	Union bolt	24	Dowel pin
7	Spring	25	Oil cooler
8	Ball	26	Grommet
9	Spring pin	27	Collar
10	O-ring	28	Washer
11	Adaptor plate	29	Bolt
12	Plunger	30	Right-hand support bracket
13	Spring		
14	Washer	31	Left-hand support bracket
15	Circlip		
16	O-ring	32	Bolt
17	Union bolt	33	Bolt
18	Plate washer		

11 Electronic ignition system: removal and fitting of special shear-head screws — late US XS850 G and SG models

1   The electronic ignition system fitted to the XS850 G (UK), XS850 G (US) and XS850 SG models is identical to the system described for the XS750 models covered in Chapter 7. When servicing and adjusting the system, follow the information given in the relevant Sections of Chapter 7 whilst noting the figures given in the Specifications Section of this Chapter.

2   It will be seen that on US versions of the G and SG models, the pick-up base plate retaining screws are of the shear-head type. Removal and fitting of these screws requires accurate use of a drill and the use of an 'easy out' screw extractor as well as a special installation tool with which to tighten the screws to a point where the heads shear off. If these tools are not readily available it may be considered advisable to place the machine in the care of an official Yamaha Service Agent who will be able to complete the job with the equipment at his disposal.

3   If it is decided to attempt the removal and fitting of these screws, commence by preparing the ends of the screws prior to drilling. This should be done by first flattening the surfaces of the screw ends with a parallel punch and then using a centre-punch to make accurately a deep centre mark in each flattened screw end. Select a drill bit of 3 mm diameter and mark a point 10 mm from its cutting end with a piece of tape so that the holes drilled are no more than 10 mm deep. Drill a hole in each screw end, taking great care to keep the drill in the centre of

the screw and thus avoiding damage to the screw thread. Clean away all traces of swarf from the base plate assembly and withdraw each screw with the 'easy out' extractor.

4   When fitting new screws, do not omit to fit the special washers beneath the screw heads and only tighten each screw after the ignition timing has been set. A special socket (tool No 90890-01308-00) should be used for tightening the screws to a point where the heads shear off.

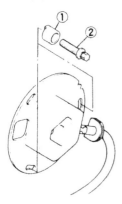

Fig. 8.8. Pick-up base plate assembly — late US XS850 G and SG models

1   Special washer   2   Shear-head bolt

12 Electronic ignition system: checking the ignition timing — XS850 H, SH and LH

1   Unlike the electronic ignition system fitted to earlier XS850 models, the system fitted to the above listed models is of the non-adjustable type. Ignition timing can be checked by following the procedure given in paragraphs 1 and 2 of Section 14, Chapter 7, whilst noting the figures given in the Specifications Section of this Chapter.

2   The '1F' mark quoted in the above mentioned text is replaced with the mark shown on the figure accompanying this text. The stationary pointer should align with the area within this mark with the engine running at 1100 rpm. Should it fail to align as stated or fail to steady, carry out a check on the ignition system components for serviceability and security, renewing or tightening any component as necessary. Do not attempt to bend the stationary pointer as a means of achieving alignment.

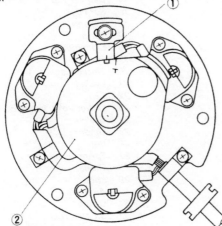

Fig. 8.9. Ignition timing adjustment — XS850 H, SH and LH

1   *Stationary pointer*   2   *Timing plate*

13 Electronic ignition system: removal and refitting of the pick-up coil and ATU assemblies

XS850 G (UK and US) and SG

1   The procedure for removing and refitting the pick-up coil assembly differs between the UK and US XS850 models only in that the UK model has normal crosshead screws securing the base plate to the crankcase cover whereas the US models have special shear-head screws. Reference should be made to Section 11 of this Chapter when removing and fitting the shear-head screws.

2   Commence removal of the pick-up coil assembly by removing the pick-up coil lead rubber guide plug from its location in the crankcase cover and detaching the lead from its retaining clips on the crankcase. Disconnect the neutral switch lead, remove the hexagon-headed bolt from the centre of the automatic timing unit (ATU) and remove the three base plate securing screws. Withdraw both the pick-up coil and ATU assemblies from the crankcase cover housing.

3   To refit the pick-up coil and ATU assemblies, reverse the above procedure whilst noting the following points. Before refitting the ATU assembly, ensure that the weights on the ATU pivot smoothly about their retaining pins and are returned to their fully retarded position by the return springs. If movement of the weights is found to be erratic or stiff, remove the clip retaining each weight to its pivot pin and pull the weight from the pin. Clean any corrosion from the pin or weight bore and lubricate each pin with a small amount of molybdenum

disulphide grease. Before refitting the pins, inspect each return spring for signs of fatigue or corrosion. If imminent failure of a spring seems likely, it is advisable to renew the ATU assembly rather than risk the spring breaking whilst the engine is in operation.

4   When refitting the ATU assembly, ensure that the pin on the crankshaft end is in alignment with the slot in the ATU shaft. When refitting the pick-up coil assembly, ensure that the slot in the base plate aligns with the corresponding projection on the ATU shaft. Fit the base plate securing screws and tighten them lightly so that the base plate is retained in position. Refit the governor assembly retaining bolt and tighten it to a torque loading of 17 lbf ft (2.3 kgf m).

5   Prior to fully tightening the base plate securing screws, set the left-hand (No 1 cylinder) piston at top dead centre (TDC) by fitting a dial test indicator assembly in the spark plug hole and rotating the crankshaft in an anti-clockwise direction until TDC is found. With the crankshaft position so set, loosen the single screw retaining the timing plate in position and move the plate so that its timing mark aligns exactly with the 'T' mark on the ATU. Tighten the screw and recheck the position of the piston at TDC and the alignment of the 'T' mark and timing plate mark.

6   Refer to Section 14 of Chapter 7 and check the ignition timing before finally tightening the pick-up base plate securing screws and relocating the pick-up coil lead in its retaining clips and guide.

XS850 H, SH and LH

7   Because the pick-up coil assembly fitted to the above listed models is of the non-adjustable type, removal and refitting of the assembly and the timing plate is greatly simplified in comparison with the procedure used on earlier models. Removal of the timing plate may be achieved simply by unscrewing and removing the allen-headed bolt holding it in position on the end of the crankshaft. To remove the pick-up coil assembly, unscrew its three retaining screws and release the pick-up coil lead rubber guide plug from its location in the crankcase cover and from its retaining clips on the crankcase.

8   To refit the pick-up coil assembly and timing plate simply reverse the above procedure, noting the locating pin on the crankshaft end and the corresponding slot in the timing plate shaft. With the pin and slot aligned and the timing plate pushed fully home in its location, tighten the retaining bolt to a torque loading of 21.5 lbf ft (3.0 kgf m).

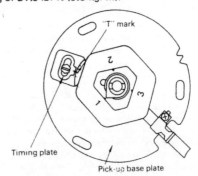

Fig. 8.10 Refitting the pick-up coil and ATU assemblies — XS850 G (UK and US) and SG

14 Front fork assembly: changing the fork oil

1   The procedure for changing the fork oil in the air assisted type of fork legs fitted to the machines covered in this Chapter is identical to that procedure given in Section 16 of Chapter 7; with the one exception that each fork leg must be depressurised prior to draining the fork oil from it and recharged with air in accordance with the instructions given in the following Section of this Chapter on completion of changing the fork oil.

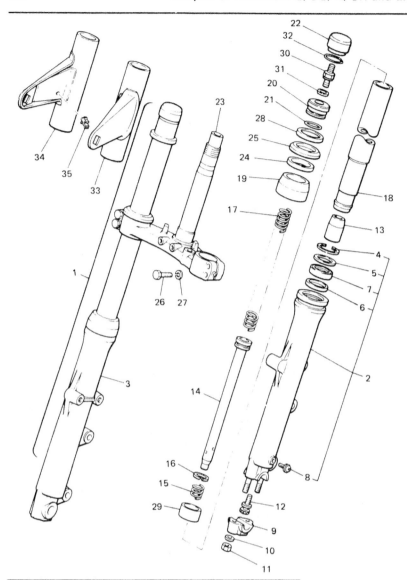

Fig. 8.11. Front fork assembly —
XS850 G (US) and H models

1   Front fork assembly
2   Left-hand lower leg
3   Right-hand lower leg
4   Clip
5   Washer
6   O-ring
7   Oil seal
8   Drain screw
9   Spindle clamp
10   Plate washer
11   Lock nut
12   Bolt
13   Oil lock piece
14   Damper rod
15   Spring
16   Piston ring
17   Spring
18   Stanchion
19   Dust seal
20   Spring seat
21   Gasket
22   Cap
23   Lower yoke assembly
24   Gasket
25   Guide
26   Bolt
27   Spring washer
28   Guide
29   Slide
30   Air valve
31   O-ring
32   Stopper ring
33   Left-hand shroud
34   Right-hand shroud
35   Clip

15 Front fork assembly: removal, dismantling, examination, renovation, reassembly and refitting — all models except XS850 G (UK)

1   The front fork legs fitted to all the XS850 models covered in this Chapter, except the XS850 G (UK), are air pressure assisted. It is essential that the air pressure contained within the fork legs is released prior to their removal and that the fork legs are recharged on completion of reassembly and refitting in accordance with the instructions given below.

2   The procedure for removing the front fork legs is similar to that given in the main text of this manual. Pressure can be released from each fork leg by removing the cap at the top of the leg to expose the charging valve and depressing the valve for one or two seconds until all air has been released.

3   To dismantle each fork leg follow the general instructions given in the main text whilst referring to the appropriate figure accompanying this text for detail changes to the assembly. Examination of each fork leg assembly should also be carried out in a method similar to that described in the main text whereas the procedure for reassembly is generally a reversal of that given for dismantling, noting the figures given in the Specifications at the beginning of this Chapter on spring free length and oil capacity and also the instructions given in the main text.

4   Note before charging each fork leg that the maximum permissible air pressure each leg is capable of accepting is 36 psi (2.5 kg/sq cm). This pressure should not be exceeded because there is a very real danger of damaging the seals within the leg.

5   If the fork legs are to be recharged with the front wheel fitted, the machine must be positioned on its main stand so that the front wheel is raised clear of the ground. Recharge the air pressure using a manually operated pump; it is not advisable to use a compressed air supply because the pressures required are very low and the volume within each fork very small. Depending on the type of handlebars fitted, the handlebars may have to be removed from the top fork yoke in order to allow the pump or air pressure gauge to be connected to the air valve.

6   Refer to the air pressure figures given in the Specifications at the beginning of this Chapter and charge each fork to the correct pressure. It is important that the pressure differential between the two fork legs does not exceed 1.4 psi (0.1 kg/sq cm) otherwise handling of the machine will be adversely affected. Do not omit to refit the cap over the air valve because this affords very necessary protection of the valve against dirt and moisture.

7   The front fork assembly fitted to the XS850 G (UK) model is identical to that shown in Chapter 7 for the later XS750 models. When servicing these forks use the procedures given in Chapter 7 whilst noting the figures given in the Specifications Section of this Chapter.

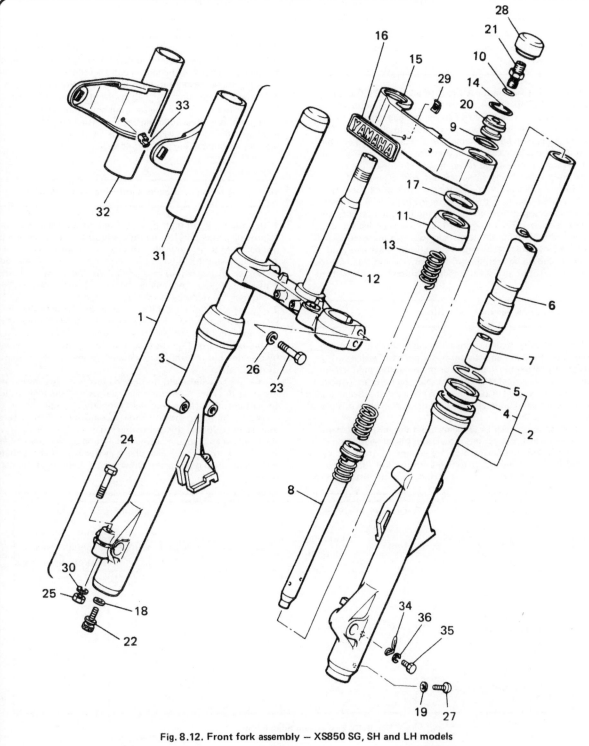

Fig. 8.12. Front fork assembly — XS850 SG, SH and LH models

1	Front fork assembly	13	Spring	25	Nut
2	Left-hand lower leg	14	Circlip	26	Spring washer
3	Right-hand lower leg	15	Cover	27	Drain screw
4	Oil seal	16	Emblem	28	Cap
5	Clip	17	Washer	29	Spring nut
6	Stanchion	18	Gasket	30	Spring washer
7	Oil lock piece	19	Gasket	31	Left-hand shroud
8	Damper rod	20	Spring seat	32	Right-hand shroud
9	O-ring	21	Air valve	33	Clip
10	O-ring	22	Bolt	34	Cable guide
11	Dust seal	23	Bolt	35	Bolt
12	Lower yoke assembly	24	Bolt	36	Spring washer

16 Rear suspension units: adjustment — all models except XS850 G (UK)

1   The rear suspension units fitted to all the XS850 models, except the XS850 G (UK), have four damping adjustment settings to compliment the five spring preload settings found as standard on the suspension units fitted to all models. The table accompanying this text shows the recommended combination of settings between the air pressure in the front fork legs and the spring setting and damping adjustment of the rear suspension units. Using these settings will ensure that the machine gives the optimum ride for the type of load carried.
2   A knurled ring near the upper mounting eye of the unit provides the adjustment control for the damping settings and is marked from 1 (soft) to 4 (hard). The spring preload setting is adjusted by turning the lower spring seat, thus altering the effective length of the spring. A screwdriver shaft or round metal rod should be used to turn the seat. The indicated position A will provide the softest ride and the position E the hardest ride.

17 Touring fairing: removal and refitting

1   Removal and refitting of the fairing assembly will be made easier with the aid of an assistant to support the fairing and thus greatly lessen the likelihood of the mounting points being subjected to abnormal stress as the mounting bolts are removed or fitted. Care should also be taken to ensure that a clean padded surface is prepared upon which to lay the fairing once it is lifted clear of the machine.
2   Commence removal by disconnecting the fairing to main harness electrical connections, situated at the left-hand side of the fairing. With the fairing properly supported, unscrew and remove the mounting bolts starting with the four bolts supporting the lower portions of the fairing. Carefully lift the fairing

clear of the machine and place it on the prepared surface.
3   To refit the fairing, position it in place on the lower support bracket and fit the retaining bolts; finger-tight only. Position the lower portions of the fairing between the main fairing and bottom rubber gasket and align them with the fairing before fitting and finally tightening all securing bolts. Care must be taken not to overtighten the securing bolts and to support the fairing properly at all times during the fitting procedure. Finally reconnect the electrical connections, referring to the wiring diagram accompanying this text as necessary.

18 Touring fairing: removing and fitting the windshield

1   It must be noted that the fairing windshield is designed only to provide protection for the rider against the elements and is not designed to give any form of protection during a collision. A cracked windshield will most certainly fail should the rider be thrown against it thus resulting in possible injury to the rider. It therefore follows that any cracked or severely scratched windshield must be replaced as soon as possible.
2   To remove the windshield, loosen and remove the nine securing screws in the reverse order to that shown in the figure accompanying this text. Once the windshield is removed it should be discarded (in the case of a cracked item) or renovated by carrying out the procedure detailed in the following Section of this Chapter.
3   To fit the windshield, loosely mount it on the fairing by inserting the nine plastic securing screws and tightening them finger-tight. With the windshield properly aligned, tighten the screws in the order shown in the accompanying figure. It is important to follow this tightening sequence in order to prevent the windscreen cracking during installation. Check tighten each screw and carry out a similar check at frequent intervals thereafter in order to prevent the screws working loose during normal usage.

	Front fork	Rear shock absorber		Loading condition		
	Air pressure	Spring seat	Damping adjuster	Solo rider	With passenger	With accessory equipments and/or passenger
1.	0.4~1.0 kg/cm² (5.7~14 psi)	A ~ E	1	○		
2.	0.4~1.0 kg/cm² (5.7~14 psi)	A ~ E	2	○	○	
3.	1.0~1.5 kg/cm² (14~21 psi)	C ~ E	3		○	○
4.	1.5 kg/cm² (21 psi)	E	4			○

Fig. 8.13. Suspension adjustment table

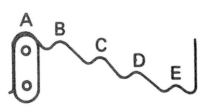

Fig. 8.14. Rear suspension unit adjuster positions

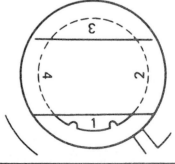

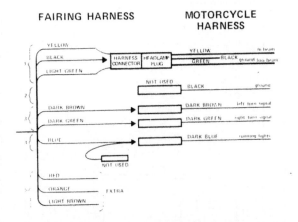

Fig. 8.15. Touring fairing to main harness electrical connections

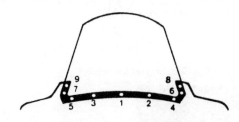

Fig. 8.16. Fairing windshield screw tightening sequence

19 Touring fairing: cleaning and renovating the windshield

1   Correct cleaning of the windshield during day-to-day usage of a machine is a part of the regular routine maintenance schedule that should be strictly adhered to as a dirty or severely scratched windshield will prevent full and clear vision and thus present an unacceptable hazard when riding the machine.
2   Always wash the windshield with a mild solution of soapy water. This solution should be liberally applied with a soft clean cloth or chamois leather; the idea being to float the dirt, dead insects, etc, from the surface of the windshield rather than rub them off with the cloth. Once the majority of contamination has been removed from the windshield by this method, stubborn traces may be removed by light rubbing with a well soaked cloth.
3   Removal of tar, oil or grease from the windshield surface can be achieved by small applications of paraffin (kerosene), naptha or isopropyl alcohol. These cleaning agents should be applied with a small piece of soft, clean cloth and applied only to the contaminated area. Directly the contamination has been removed by this method, rinse the affected area once more with a solution of soapy water to remove any remaining oily traces.
4   Paraffin (kerosene) naptha and isopropyl alcohol are the only three cleaning agents recommended by Yamaha; in no circumstances should petrol (gasolene), ethyl or methly alcohol, abrasive cleaners, cleaner wax combinations or any form of solvent be used because all these agents will adversely affect the plastic from which the windshield is constructed. Take great care when using these recommended cleaning agents to wear adequate protective clothing, especially over the eyes. Always carry out the cleaning procedure in a well ventilated area, well clear of any source of sparks or naked flame.
5   If the windshield has been scratched but the scratches do not extend to any great depth, it is worth attempting to remove the

scratches by carrying out the following procedure rather than have to discard the windshield.
6   Each scratch should first be removed by gentle application of 400 – 600 grade wet-and-dry paper to the immediate area. Once the line of the scratch has disappeared, the area should be polished with a high quality plastic polishing compound such as jewellers' rouge applied to either a soft, clean cloth or a clean soft-muslin buffing wheel.

20 Pannier assembly: removal and refitting

1   On models fitted with panniers as standard equipment, it is necessary to remove the pannier assembly in order to carry out rear wheel removal. Before removal of the pannier assembly can be started, it is first necessary to detach the seat from its frame attachment points and remove the luggage carrier by first unscrewing its mounting point attachment bolts and then pulling the assembly rearwards to clear the machine.
2   Commence removal of the pannier assembly by disconnecting the two electrical leads leading to the rear flashing indicators. Remove the pannier guard rail pinch bolts at the frame guards and remove the pannier mounting bolts that secure the panniers to the frame guard bracket. Move to the rear of the frame and remove the mounting plate bolts; these being the bolts that attach to the old indicator mountings. Remove the side plates from the rear suspension unit mounting studs and lift the complete assembly from the rear of the machine.
3   When refitting the pannier assembly, ensure that the side plates are correctly located over the rear suspension unit mounting studs. Refit the mounting plate bolts at the rear of the frame and secure the panniers to the frame guard brackets. Insert the guard rails into the split-tubes of the frame guards and refit and tighten the pinch bolts. Finally, reconnect the electrical leads to the flashing indicators and check that each indicator functions correctly before refitting the luggage carrier and seat assemblies. Before riding the machine, ensure that no part of the pannier assembly comes into contact with any moving rear suspension component.

21 Starting circuit cut-off system: general description — XS850 H, SH and LH

1   The above listed models are equipped with a system whereby the starter motor will not function unless neutral gear is selected before the starter button is pressed. Alternatively, the machine may be left in gear but the clutch lever pulled in to disengage the clutch before pressing the starter button.
2   A small switch fitted to the right-hand side of the crankcase operates a warning lamp in the instrument panel to indicate that neutral has been selected. More importantly, it is interconnected with the starter solenoid and will only allow the engine to be started if the gearbox is in neutral, unless the clutch is disengaged. It can be checked by setting a multimeter on the resistance scale and connecting one probe to the switch terminal and the other to earth. The meter should indicate continuity when neutral is selected and infinite resistance when in any gear. This switch can be difficult to locate because it is well hidden beneath a small plate which itself is partially hidden by the right-hand exhaust pipe just forward of its union with the middle pipe. The switch can be exposed by removing the single Allen-headed screw retaining the plate in position.
3   A small plunger-type switch is incorporated in the clutch lever, serving to prevent operation of the starter circuit when any gear has been selected, unless the clutch lever is held in. It can be checked by the method described above for the neutral switch. If defective it must be renewed, as there is no satisfactory means of repair.

186

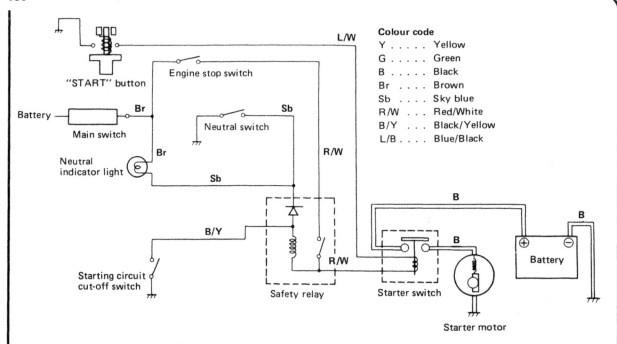

Fig. 8.17. Starting circuit cut-off system circuit diagram — XS850 H, SH and LH models

Colour code
Y . . . . . Yellow
G . . . . . Green
B . . . . Black
Br . . . Brown
Sb . . . Sky blue
R/W . . . Red/White
B/Y . . . Black/Yellow
L/B . . . . Blue/Black

21.2 Remove the cover plate to expose the neutral indicator switch

Colour code for Yamaha XS850 wiring diagrams

R	Red	G	Green
B	Black	W	White
Sb	Sky blue	P	Pink
Br	Brown	Gy	Gray
Ch	Chocolate	O	Orange
Dg	Dark green	B/W	Black/White
L	Blue	G/Y	Green/Yellow
Y	Yellow	Y/W	Yellow/White
Lg	Light green	L/W	Blue/White
Y/R	Yellow/Red	Br/W	Brown/White
L/B	Blue/Black	L/Y	Blue/Yellow
B/R	Black/Red	R/W	Red/White
W/G	White/Green	L/G	Blue/Green

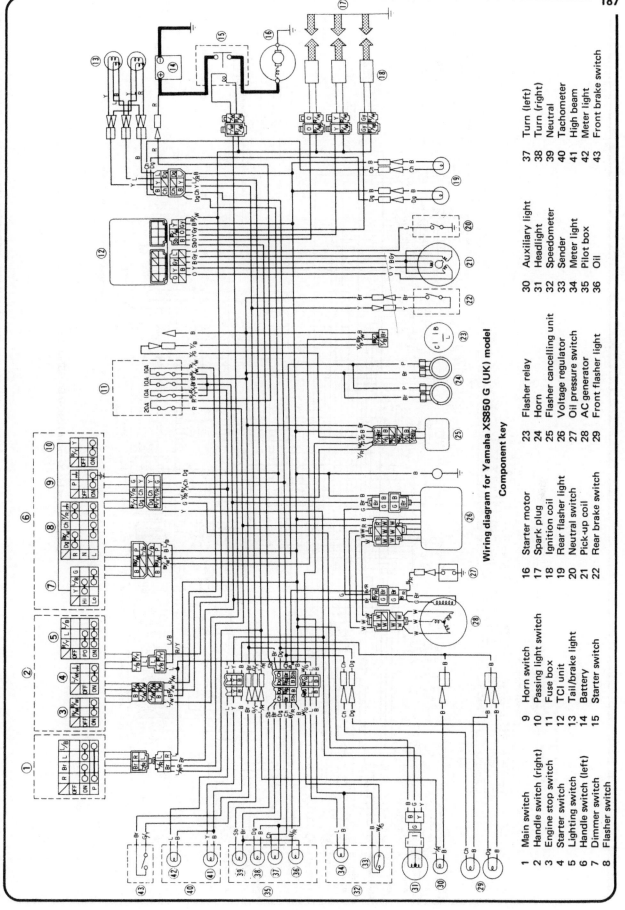

Wiring diagram for Yamaha XS850 G (UK) model

Component key

1	Main switch	16	Starter motor	30	Auxiliary light	37	Turn (left)
2	Handle switch (right)	17	Spark plug	31	Headlight	38	Turn (right)
3	Engine stop switch	18	Ignition coil	32	Speedometer	39	Neutral
4	Starter switch	19	Rear flasher light	33	Sender	40	Tachometer
5	Lighting switch	20	Neutral switch	34	Meter light	41	High beam
6	Handle switch (left)	21	Pick-up coil	35	Pilot box	42	Meter light
7	Dimmer switch	22	Rear brake switch	36	Oil	43	Front brake switch
8	Flasher switch	23	Flasher relay				
9	Horn switch	24	Horn				
10	Passing light switch	25	Flasher cancelling unit				
11	Fuse box	26	Voltage regulator				
12	TCI unit	27	Oil pressure switch				
13	Tail/brake light	28	AC generator				
14	Battery	29	Front flasher light				
15	Starter switch						

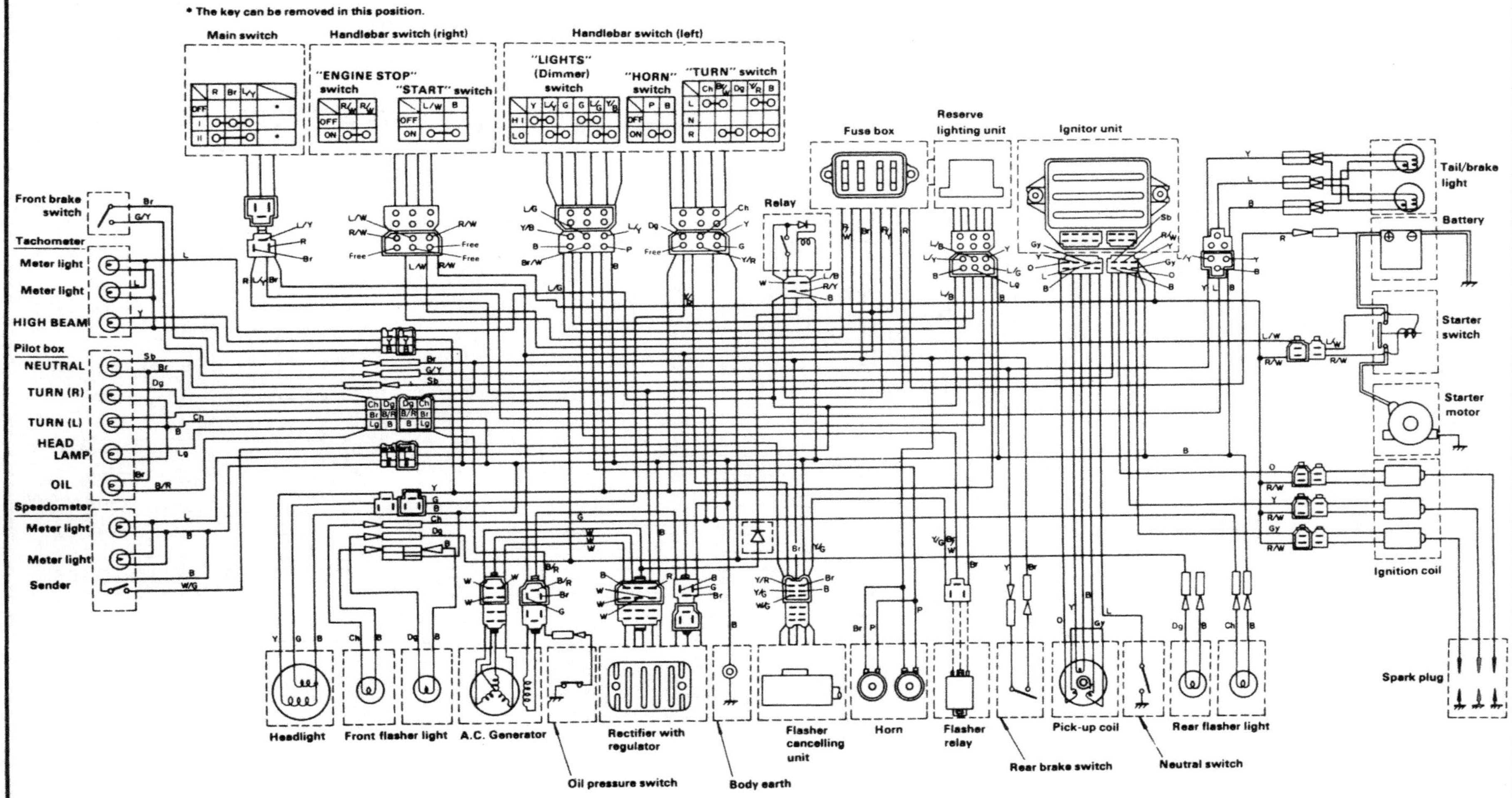

* The key can be removed in this position.

Wiring diagram for Yamaha XS850 G (US) model

Tail/brake light
Battery
Starter switch
Starter motor
Ignition coil
Spark plug

Rear flasher light
Neutral switch
Pick-up coil
Rear brake switch
Flasher relay
Horn
Flasher cancelling unit
Body earth
Rectifier with regulator
Oil pressure switch
A.C. Generator
Front flasher light
Headlight

Ignitor unit
Reserve lighting unit
Fuse box
Relay

Handlebar switch (left)
"TURN" switch
"HORN" switch
"LIGHTS" (Dimmer) switch

Handlebar switch (right)
"ENGINE STOP" switch
"START" switch

Main switch

	R	Br	L/W
OFF			
–			
–			

Front brake switch
Tachometer
Meter light
Meter light
HIGH BEAM
Pilot box
NEUTRAL
TURN (R)
TURN (L)
HEAD LAMP
OIL
Speedometer
Meter light
Meter light
Sender

189

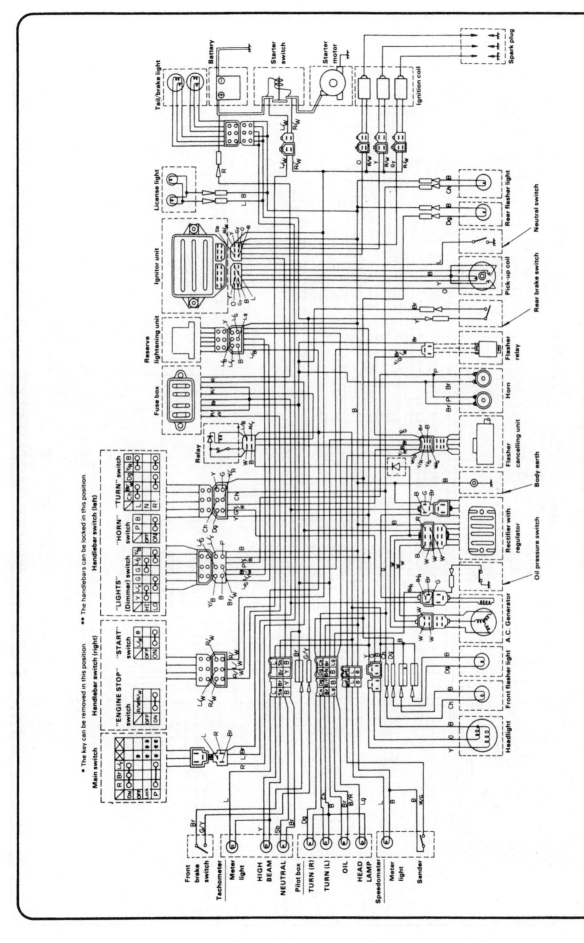

Wiring diagram for Yamaha XS850 SG model

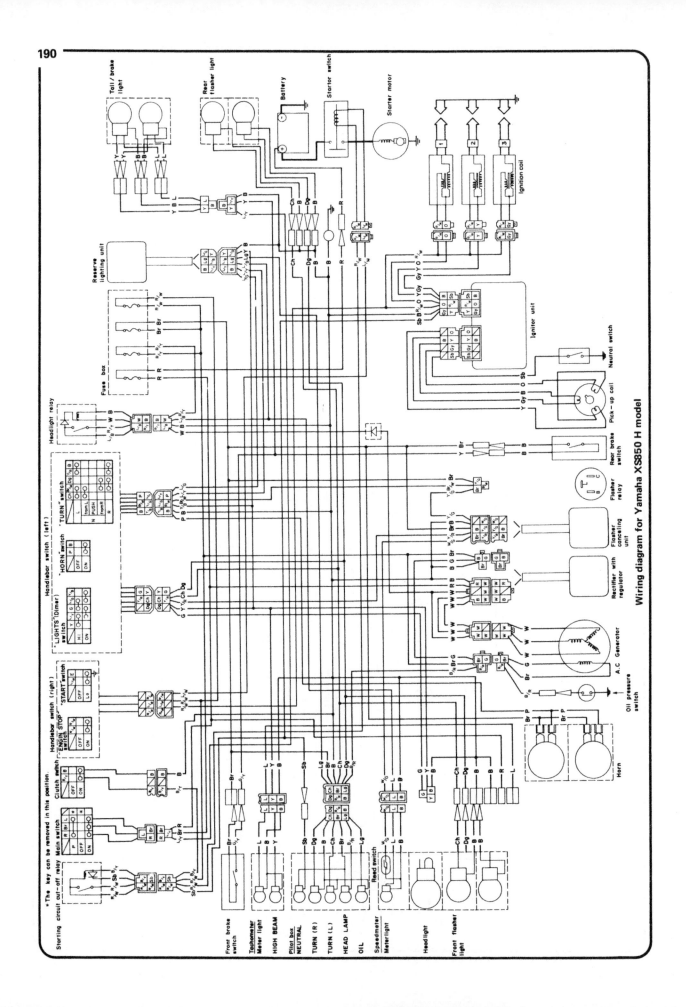

Wiring diagram for Yamaha XS850 H model

Wiring diagram for Yamaha XS850 SH and LH models

Conversion factors

Length (distance)
Inches (in)	X	25.4	= Millimetres (mm)	X 0.039	= Inches (in)
Feet (ft)	X	0.305	= Metres (m)	X 3.281	= Feet (ft)
Miles	X	1.609	= Kilometres (km)	X 0.621	= Miles

Volume (capacity)
Cubic inches (cu in; in³)	X	16.387	= Cubic centimetres (cc; cm³)	X 0.061	= Cubic inches (cu in; in³)
Imperial pints (Imp pt)	X	0.568	= Litres (l)	X 1.76	= Imperial pints (Imp pt)
Imperial quarts (Imp qt)	X	1.137	= Litres (l)	X 0.88	= Imperial quarts (Imp qt)
Imperial quarts (Imp qt)	X	1.201	= US quarts (US qt)	X 0.833	= Imperial quarts (Imp qt)
US quarts (US qt)	X	0.946	= Litres (l)	X 1.057	= US quarts (US qt)
Imperial gallons (Imp gal)	X	4.546	= Litres (l)	X 0.22	= Imperial gallons (Imp gal)
Imperial gallons (Imp gal)	X	1.201	= US gallons (US gal)	X 0.833	= Imperial gallons (Imp gal)
US gallons (US gal)	X	3.785	= Litres (l)	X 0.264	= US gallons (US gal)

Mass (weight)
Ounces (oz)	X	28.35	= Grams (g)	X 0.035	= Ounces (oz)
Pounds (lb)	X	0.454	= Kilograms (kg)	X 2.205	= Pounds (lb)

Force
Ounces-force (ozf; oz)	X	0.278	= Newtons (N)	X 3.6	= Ounces-force (ozf; oz)
Pounds-force (lbf; lb)	X	4.448	= Newtons (N)	X 0.225	= Pounds-force (lbf; lb)
Newtons (N)	X	0.1	= Kilograms-force (kgf; kg)	X 9.81	= Newtons (N)

Pressure
Pounds-force per square inch (psi; lbf/in²; lb/in²)	X	0.070	= Kilograms-force per square centimetre (kgf/cm²; kg/cm²)	X 14.223	= Pounds-force per square inch (psi; lbf/in²; lb/in²)
Pounds-force per square inch (psi; lbf/in²; lb/in²)	X	0.068	= Atmospheres (atm)	X 14.696	= Pounds-force per square inch (psi; lbf/in²; lb/in²)
Pounds-force per square inch (psi; lbf/in²; lb/in²)	X	0.069	= Bars	X 14.5	= Pounds-force per square inch (psi; lbf/in²; lb/in²)
Pounds-force per square inch (psi; lbf/in²; lb/in²)	X	6.895	= Kilopascals (kPa)	X 0.145	= Pounds-force per square inch (psi; lbf/in²; lb/in²)
Kilopascals (kPa)	X	0.01	= Kilograms-force per square centimetre (kgf/cm²; kg/cm²)	X 98.1	= Kilopascals (kPa)

Torque (moment of force)
Pounds-force inches (lbf in; lb in)	X	1.152	= Kilograms-force centimetre (kgf cm; kg cm)	X 0.868	= Pounds-force inches (lbf in; lb in)
Pounds-force inches (lbf in; lb in)	X	0.113	= Newton metres (Nm)	X 8.85	= Pounds-force inches (lbf in; lb in)
Pounds-force inches (lbf in; lb in)	X	0.083	= Pounds-force feet (lbf ft; lb ft)	X 12	= Pounds-force inches (lbf in; lb in)
Pounds-force feet (lbf ft; lb ft)	X	0.138	= Kilograms-force metres (kgf m; kg m)	X 7.233	= Pounds-force feet (lbf ft; lb ft)
Pounds-force feet (lbf ft; lb ft)	X	1.356	= Newton metres (Nm)	X 0.738	= Pounds-force feet (lbf ft; lb ft)
Newton metres (Nm)	X	0.102	= Kilograms-force metres (kgf m; kg m)	X 9.804	= Newton metres (Nm)

Power
Horsepower (hp)	X	745.7	= Watts (W)	X 0.0013	= Horsepower (hp)

Velocity (speed)
Miles per hour (miles/hr; mph)	X	1.609	= Kilometres per hour (km/hr; kph)	X 0.621	= Miles per hour (miles/hr; mph)

Fuel consumption*
Miles per gallon, Imperial (mpg)	X	0.354	= Kilometres per litre (km/l)	X 2.825	= Miles per gallon, Imperial (mpg)
Miles per gallon, US (mpg)	X	0.425	= Kilometres per litre (km/l)	X 2.352	= Miles per gallon, US (mpg)

Temperature

Degrees Fahrenheit (°F) $= (°C \times \frac{9}{5}) + 32$

Degrees Celsius (Degrees Centigrade; °C) $= (°F - 32) \times \frac{5}{9}$

*It is common practice to convert from miles per gallon (mpg) to litres/100 kilometres (l/100km), where mpg (Imperial) x l/100 km = 282 and mpg (US) x l/100 km = 235

Metric conversion tables

Inches	Decimals	Millimetres	Millimetres to Inches		Inches to Millimetres	
			mm	Inches	Inches	mm
1/64	0.015625	0.3969	0.01	0.00039	0.001	0.0254
1/32	0.03125	0.7937	0.02	0.00079	0.002	0.0508
3/64	0.046875	1.1906	0.03	0.00118	0.003	0.0762
1/16	0.0625	1.5875	0.04	0.00157	0.004	0.1016
5/64	0.078125	1.9844	0.05	0.00197	0.005	0.1270
3/32	0.09375	2.3812	0.06	0.00236	0.006	0.1524
7/64	0.109375	2.7781	0.07	0.00276	0.007	0.1778
1/8	0.125	3.1750	0.08	0.00315	0.008	0.2032
9/64	0.140625	3.5719	0.09	0.00354	0.009	0.2286
5/32	0.15625	3.9687	0.1	0.00394	0.01	0.254
11/64	0.171875	4.3656	0.2	0.00787	0.02	0.508
3/16	0.1875	4.7625	0.3	0.01181	0.03	0.762
13/64	0.203125	5.1594	0.4	0.01575	0.04	1.016
7/32	0.21875	5.5562	0.5	0.01969	0.05	1.270
15/64	0.234375	5.9531	0.6	0.02362	0.06	1.524
1/4	0.25	6.3500	0.7	0.02756	0.07	1.778
17/64	0.265625	6.7469	0.8	0.03150	0.08	2.032
9/32	0.28125	7.1437	0.9	0.03543	0.09	2.286
19/64	0.296875	7.5406	1	0.03937	0.1	2.54
5/16	0.3125	7.9375	2	0.07874	0.2	5.08
21/64	0.328125	8.3344	3	0.11811	0.3	7.62
11/32	0.34375	8.7312	4	0.15748	0.4	10.16
23/64	0.359375	9.1281	5	0.19685	0.5	12.70
3/8	0.375	9.5250	6	0.23622	0.6	15.24
25/64	0.390625	9.9219	7	0.27559	0.7	17.78
13/32	0.40625	10.3187	8	0.31496	0.8	20.32
27/64	0.421875	10.7156	9	0.35433	0.9	22.86
7/16	0.4375	11.1125	10	0.39370	1	25.4
29/64	0.453125	11.5094	11	0.43307	2	50.8
15/32	0.46875	11.9062	12	0.47244	3	76.2
31/64	0.48375	12.3031	13	0.51181	4	101.6
1/2	0.5	12.7000	14	0.55118	5	127.0
33/64	0.515625	13.0969	15	0.59055	6	152.4
17/32	0.53125	13.4937	16	0.62992	7	177.8
35/64	0.546875	13.8906	17	0.66929	8	203.2
9/16	0.5625	14.2875	18	0.70866	9	228.6
37/64	0.578125	14.6844	19	0.74803	10	254.0
19/32	0.59375	15.0812	20	0.78740	11	279.4
39/64	0.609375	15.4781	21	0.82677	12	304.8
5/8	0.625	15.8750	22	0.86614	13	330.2
41/64	0.640625	16.2719	23	0.90551	14	355.6
21/32	0.65625	16.6687	24	0.94488	15	381.0
43/64	0.671875	17.0656	25	0.98425	16	406.4
11/16	0.6875	17.4625	26	1.02362	17	431.8
45/64	0.703125	17.8594	27	1.06299	18	457.2
23/32	0.71875	18.2562	28.	1.10236	19	482.6
47/64	0.734375	18.6531	29	1.14173	20	508.0
3/4	0.75	19.0500	30	1.18110	21	533.4
49/64	0.765625	19.4469	31	1.22047	22	558.8
25/32	0.78125	19.8437	32	1.25984	23	584.2
51/64	0.796875	20.2406	33	1.29921	24	609.6
13/16	0.8125	20.6375	34	1.33858	25	635.0
53/64	0.828125	21.0344	35	1.37795	26	660.4
27/32	0.84375	21.4312	36	1.41732	27	685.8
55/64	0.859375	21.8281	37	1.4567	28	711.2
7/8	0.875	22.2250	38	1.4961	29	736.6
57/64	0.890625	22.6219	39	1.5354	30	762.0
29/32	0.90625	23.0187	40	1.5748	31	787.4
59/64	0.921875	23.4156	41	1.6142	32	812.8
15/16	0.9375	23.8125	42	1.6535	33	838.2
61/64	0.953125	24.2094	43	1.6929	34	863.6
31/32	0.96875	24.6062	44	1.7323	35	889.0
63/64	0.984375	25.0031	45	1.7717	36	914.4

English/American terminology

Because this book has been written in England, British English component names, phrases and spellings have been used throughout. American English usage is quite often different and whereas normally no confusion should occur, a list of equivalent terminology is given below.

English	American	English	American
Air filter	Air cleaner	Number plate	License plate
Alignment (headlamp)	Aim	Output or layshaft	Countershaft
Allen screw/key	Socket screw/wrench	Panniers	Side cases
Anticlockwise	Counterclockwise	Paraffin	Kerosene
Bottom/top gear	Low/high gear	Petrol	Gasoline
Bottom/top yoke	Bottom/top triple clamp	Petrol/fuel tank	Gas tank
Bush	Bushing	Pinking	Pinging
Carburettor	Carburetor	Rear suspension unit	Rear shock absorber
Catch	Latch	Rocker cover	Valve cover
Circlip	Snap ring	Selector	Shifter
Clutch drum	Clutch housing	Self-locking pliers	Vise-grips
Dip switch	Dimmer switch	Side or parking lamp	Parking or auxiliary light
Disulphide	Disulfide	Side or prop stand	Kick stand
Dynamo	DC generator	Silencer	Muffler
Earth	Ground	Spanner	Wrench
End float	End play	Split pin	Cotter pin
Engineer's blue	Machinist's dye	Stanchion	Tube
Exhaust pipe	Header	Sulphuric	Sulfuric
Fault diagnosis	Trouble shooting	Sump	Oil pan
Float chamber	Float bowl	Swinging arm	Swingarm
Footrest	Footpeg	Tab washer	Lock washer
Fuel/petrol tap	Petcock	Top box	Trunk
Gaiter	Boot	Torch	Flashlight
Gearbox	Transmission	Two/four stroke	Two/four cycle
Gearchange	Shift	Tyre	Tire
Gudgeon pin	Wrist/piston pin	Valve collar	Valve retainer
Indicator	Turn signal	Valve collets	Valve cotters
Inlet	Intake	Vice	Vise
Input shaft or mainshaft	Mainshaft	Wheel spindle	Axle
Kickstart	Kickstarter	White spirit	Stoddard solvent
Lower leg	Slider	Windscreen	Windshield
Mudguard	Fender		

Index